Seeds, Sex and Civilization

Seeds, Sex and Civilization

How the Hidden Life of Plants has Shaped our World

Peter Thompson

With 49 illustrations, 43 in color

Thames & Hudson

Publisher's Note

Peter Thompson died in 2008 during the preparation for publication of his book. At the request of his family, Dr Stephen Harris, Druce Curator of the Oxford University Herbaria, was invited to write a new conclusion to the book and to assist the publisher with seeing the book through the press. The author's family, and the publisher, warmly thank Dr Harris for his contribution.

Throughout *Seeds, Sex and Civilization*, plants have – with very few exceptions – been referred to by their vernacular names. Readers seeking to identify precisely which plants are being discussed will find the Latin equivalents of vernacular names in the index.

Frontispiece Sunflower seeds sprouting

First published in 2010 in hardcover in the United States of America by Thames & Hudson Inc., 500 Fifth Avenue, New York, New York 10110

thamesandhudsonusa.com

Library of Congress Catalog Card Number 2010923359

ISBN 978-0-500-25170-6

Printed and bound in Singapore by Tien Wah Press (Pte) Ltd

Contents

Introduction

'I have great faith in a seed…Convince me that you have a seed there, and I am prepared to expect wonders.'

Henry D. Thoreau, *The Succession of Forest Trees*, 1860

'Seeds are for gardeners' – several people I spoke to thought that. 'Seeds are about natural history: without them there would be no wildflowers' was another feeling widely held. Seeds are actually important for us all, but hardly anyone I talked to about this book saw that gardening and natural history were cameo roles compared to the parts seeds play in our daily lives – parts so central to our existence that without them such lives would not exist. We would not inhabit civilized societies; most of us would not be alive at all. Seeds are so much a part of our lives that we take them for granted, and do not even notice they are there.

Seeds sustain us not only in foods (from cocoa beans and pulses to cereals and cooking oil), but also in fabrics, detergents and fuels. As well as these direct encounters, seeds are vital to efficient arable farming, vegetable gardening and stock raising. Leys of grass and clover grown from seeds have replaced permanent pastures in many parts of the world, and chickens, pigs and beef cattle are raised in intensive systems almost totally dependent on grains of one kind or another. Such large-scale agribusiness underpins modern economies, and the consumer lifestyles that they provide.

The long and fascinating story of humankind's association with seeds, which began more than ten thousand years ago, is the subject of this book.

Within a few thousand years, people in half a dozen widely separated parts of the world had learned how to till the land, sow seeds of grasses and other plants previously gathered from the wild, cultivate the plants that grew from them and harvest and store the seeds they produced. In the centuries that followed cultivated crops spread from these few original centres across the globe. In one continent or another seeds of wheat, barley, lentils and other pulses, rice, soybeans, sorghum, millets, maize, beans, quinoa and amaranth became staple foods for millions of people.

The nomadic lifestyles of the hunter gatherers were replaced by settled communities of farmers. The ease with which seeds could be transported, together with their ready availability and long storage lives, released people from dependence on what could be gathered, dug up or hunted day by day. Villages grew into towns in which merchants, artisans, tradesmen, shop-keepers and others went about their business, supported by the products of surrounding farms. City sites were chosen for their suitability for centres of power or trade, often at some distance from the fields of the farmers who fed their inhabitants.

These arrangements persisted virtually unchanged until relatively recently. For thousands of years peasant farmers sowed, cultivated, reaped and gathered the grain of the crops they grew, putting aside the seed corn year by year to provide for the following year's crop. Drought and frost, insect pests, fungal diseases and the climate affected plants differently from one place to another and, wherever crops were grown, different combinations of one sort with another could respond to the situation. The agricultural production of the world was gathered year by year from an intricately mixed and variable medley of different strains, known as landraces. Some landraces were more productive, others less so – but they were almost incomparably low yielding when set against today's expectations.

Over the last three hundred years, the pace of change accelerated. The startling discovery of sex in plants during the eighteenth century opened the eyes of enlightened farmers, landowners and gardeners a hundred years later to the possibilities of breeding new and improved varieties of fruits, vegetables and crop plants. Twentieth-century plant breeders drew upon the unravelling of heredity's secrets to transform the basis of agricultural production through high yielding pedigree strains. These began to replace the old landraces, at first slowly and then in an all-consuming flood, financed by governments and the major seed producers. With the loss of the old landraces went the genetic diversity built up over thousands of years of unofficial procreation in peasant farmers' fields.

Seeds, Sex and Civilization tells the story of humankind's relationship with, and dependence on, seeds. It explores how, like reluctant pupils, we gradually learned what seeds are, where they came from and their role in the survival of wildflowers and the productivity of crops. It shows how we used that knowledge to grow bigger and better plants in gardens and on farms, and how this was to prove a double-edged sword. The book reveals how hubris brought us to the brink of losing our inheritance of genetic resistance to disease, drought tolerance and other qualities on which the success of future plant breeding – and the survival of civilized societies – depends. And, finally, it traces our response to that threat – a response almost unknown, at least until recently, to most people. Yet it was to involve the creation of a worldwide network of seed banks and plant breeding establishments in one of the oldest, and certainly by far the most extensive and costly, conservation projects ever undertaken.

CHAPTER ONE

The Roots of Agriculture

Ribs of rock protruded through sun-baked, steeply sloping meadows filled with wildflowers. Crimson poppies and scarlet buttercups, the sumptuous blooms of Juno irises, multicoloured drifts of anemones, orange marigolds, chrome yellow chrysanthemums, purple lavenders and the silken pink chalices of bindweed tempted us to stop and wander through these meadows. Behind us lay the productive, cultivated fields of Galilee. We were in wilder country now – a countryside where barbed wire strung along posts around enclosures discouraged inclinations to explore. Sinister little signs with the word '*Minen*' stencilled beneath a skull and crossbones brought the message home. Three years earlier, in 1967, the Israeli army had stormed the Golan Heights, driven out the Syrians and pursued them halfway down the road to Damascus. Now nature had taken over abandoned fields around deserted towns and villages. The battlefield had become a resting place for thousands of birds migrating north towards Turkey and their summer homes in Europe. Flocks of storks snoozed among the land mines, balancing on one leg within the barbed-wire enclosures. Their long bills nestled among their breast feathers, unaware of the lethal crop beneath their feet.

We looked back from the crest of the Golan Heights over the Sea of Galilee across the land we had just left (1). Above us Mount Hermon towered towards the sky, its upper slopes snow-covered and its peaks hidden by clouds. Leaving the car, we walked through abandoned fields. Grasses with narrow, skimpy leaves and strange, spiky flower heads brushed against our knees. These were goat grasses and, despite the more colourful attractions of other wildflowers, these were what we were looking for. Although they were unpromising in appearance, we sought out the seeds of goat grasses because of their importance in the origin of bread wheat, and because the genes that they contained might prove useful in wheat breeding.

Humble goat grasses have played crucial parts in human history. For tens of thousands of years semi-nomadic tribes, living in the foothills of mountains stretching across the Middle East to Iran and beyond, had been members of the goat grass appreciation society. They had walked, just as we had that day, through natural meadows of goat grasses, intermingled with wild wheats and barleys. Using bone sickles inset with flints, they would have harvested handfuls and laid them down to dry for a few days before beating out the seeds with simple wooden flails.

Nomadic tribespeople who passed through these meadows some nineteen thousand years ago left faint traces of their activities – the earliest known remains of the roots of western agriculture. They were here in the final one thousand years of a period when it is assumed that men hunted the animals living around them and women gathered the seeds, roots and fruits of wild plants, according to the ways of their ancestors. Such tribes would have observed grasses and other useful plants spreading into patches of ground created after fires had destroyed trees and shrubs, and may have started fires themselves to encourage the process. Beyond that they relied on what nature provided. Before another ten thousand years passed their distant offspring would have taken the first steps towards cultivating plants, rather than gathering them haphazardly. In doing so, they stepped innocently into the mouth of a decoy – a decoy that would inexorably transform their descendants slowly and subtly, millennia by millennia, from a people free to travel wherever the wildflowers led them into farming communities, bound to the soil and subservient to the needs of the plants they believed they had mastered.

The grains of goat grasses are about the size of small barleycorns. They are nutritious and good to eat, but each plant produces only a few and extracting enough to make a square meal is a time-consuming, fiddly task. The skinny ears of wild wheats and barleys, containing numerous small, round grains, are more rewarding, but because the ears disintegrate as they ripen much of the harvest is lost in the gathering. The edible grains must be separated from the harsh husks by vigorous threshing and rolling. After that, the more retentive bits and pieces are burnt off in the embers of fires – a process known as parching. Some grains emerge barely singed, others are toasted brown and, when attention slips, others are burnt to charcoal. Charred grains are almost impervious to decay.

Such charred remains have been meticulously sifted from the residues of ancient campfires. They tell us what our ancestors were gathering, and later growing, as food in the Middle East, tens of thousands of years after it was harvested. These remains tell us that as glaciers relaxed their grip on northern Europe, about ten thousand years ago, the people who roamed the rocky slopes above the Sea of Galilee lived in small family groups, hunting animals and gathering what plants they could. During the next few thousand years the inhabitants of enormous swathes of territory, from southeastern Europe and the western Mediterranean to Anatolia and northern Iraq, gradually, imperceptibly, started to depart from the ways of their ancestors.

They continued to live lightly on the land, hunting game and supplementing the meat with whatever seeds and other natural products they could most easily gather. However, the mute evidence of further charred grains retrieved from ancient settlements suggests that by around twelve thousand years ago deliberate efforts were being made at Jericho, as in a number of other places, to encourage some of the more favoured grasses and pulses, if not to cultivate them. Particular kinds of grain began to be predominant in diets. In some places this was einkorn (2, 3), one of the original wild wheat species. In others it was barley, or sometimes emmer (4), a hybrid between a species of wild wheat and one of the goat grasses. An increasing tendency for one kind to be favoured over another – perhaps by deliberately encouraging it to grow where it would be more readily accessible – is indicated. The gatherers were beginning to depend less on what nature provided and to rely rather more on their own efforts. They had moved a little deeper into the decoy.

The retreat of the glaciers marks the beginning of a period known as the Mesolithic, or 'Middle Stone Age'. This period of human history ended as humans began to cultivate land during the Neolithic, or 'New Stone Age'.

Slowly, patchily and erratically, the evidence for increasing dependence on cultivated cereals accumulates over the next four thousand years. Seeds extracted laboriously and meticulously from among the charred cereal grains are less likely to be random assortments of steppe species. There is evidence of weeds – the inevitable camp followers of cultivation – including wild oats and brome grasses, fumitories, vetches, mallows, cornflowers, pheasants' eyes, marigolds, poppies and knotweeds, still all too familiar to farmers and gardeners. Later, as Neolithic farmers moved into North Africa and through Asia Minor into Europe, the seeds of such weeds accompanied them as unwillingly tolerated hitch-hikers.

The main changes that swung the balance from dependence on the collection of wild cereals to some kind of a farming economy seem to have taken place between ten thousand and seven thousand years ago. As with many historical processes, this extended period of flux is best identified with hindsight, since the beginning and end of the period will not be clearly demarcated. Furthermore, the period will have been longer or shorter depending on the places in which people lived. During this period long established ways of life were abandoned and replaced by new ways of making a living and organizing society. As the period started, people were already beginning to build more or less permanent settlements with substantial rectilinear buildings, rather than the round huts of nomadic societies. They were beginning to domesticate goats, followed shortly by sheep, foregoing some of the joys and insecurities of hunting wild game aided by dogs, their most ancient allies, for more mundane responsibilities of animal husbandry. Flocks of sheep and goats might have been critical additions to the equation that enabled man to become farmer as well as pastoralist. Goats browsed and reduced the scrub, opening the way for colonization by grasses. Sheep grazed and encouraged the formation of swards of closely cropped plants dominated by grasses. The activities of both animals would have increased the extent and fertility of the natural meadows close to settlements, making them more extensive, more productive and easier to harvest.

Previously our ancestors had gathered seeds and roots from wild meadows, picked fruit from trees and then moved on. Now they were staying longer in the places where they found useful plants most accessible. The hybrid wheat, emmer, becomes a more consistent presence in the debris of settlements spread throughout the Middle East, and weed seeds become regular associates of the charred cereal grains. People settled down. They began to scatter seeds of their preferred grains close to their settlements, and gathered them, together with wild grasses, when they ripened in spring. People had stores of seeds on which they could live throughout the year.

Then, as now, birds claimed their share of any cereals going. Small-grained crops, including barley, rice, rye, sorghum and millet, are so attractive to birds and vulnerable to their depredations that it is almost impossible to produce a worthwhile, or indeed any, crop from a small, isolated patch. On an agricultural scale this situation is avoided as a matter of course, but it is one with which the first farmers would have had to contend. Wild forms of wheat and barley are less susceptible to birds than semi-domesticated varieties. Their grains are more tightly enclosed by parts of the ear, and the brittleness of the ear itself causes it to shatter when birds start work on it. This scatters the grains on the ground, where some at least escape being eaten. In semi-domesticated varieties, more like those of today, the ear remains intact, rendering them highly vulnerable to plundering by sparrows and finches. So did our ancestors' children spend an inordinate time bird scaring in fields of ripening grain? Or were these crops already being grown on a sufficient scale to outstrip the birds' appetites by the time improved kinds of wheat became available? Were farmers, even soon after the dawn of agriculture, forced to cooperate in common field systems?

Over thousands of years, as their qualities became recognized, the higher yielding, domesticated forms of emmer and other cereals began to spread away from the foothills of the mountains. They were taken down the Tigris and Euphrates rivers into Mesopotamia and other more distant places. Our ancestors were now deep into the decoy. They were unwittingly led on by the promise of more abundant supplies of more palatable food, and more and better opportunities for barter with neighbours.

Across the Middle East – in a great arc from Mount Hebron across southern Anatolia and Iraq to northern Iran, Tajikistan and east towards the borders of China – the inhabitants of the lower mountain slopes were ideally placed to domesticate grasses growing around them. Mountains tempered the hot, dry summers experienced by lower lying, less favoured parts of this region and attracted regular, abundant rainfall in winter. Seeds of wheat, barley, rye and lentils, sown with the return of autumn rains, rapidly generated seedlings that survived the winter, then grew away strongly in spring. They produced a harvest as summer droughts turned the landscape into the semblance of a desert.

Different peoples in different places domesticated different plants depending on what they found around them and what growing conditions they faced. The peoples around the eastern Mediterranean, living in not dissimilar conditions, enjoyed the benefit of benign winters which enabled plants to grow more or less continuously, if only slowly, from autumn to spring. Five thousand miles away, hunter gatherers in northern China learned to cultivate millets, sorghum and beans in a country where freezing Siberian winds dried the landscape and its plants to a crisp from late autumn and through the winter. For them spring was the season of renewal, when seeds were sown and seedlings emerged and grew vigorously, watered by the summer monsoon carried from the Pacific Ocean. Over almost the same period, people in Yunnan domesticated buckwheat and barley on the mountainsides overlooking sub-tropical valleys. And, most astonishing of all, within a few thousand years the inhabitants of valleys in the mountains of Mexico, across the vast expanse of the Pacific Ocean, discovered the secrets of growing maize, beans and squashes and adapted them to their climate. They harvested their seeds towards the end of hot, moist summers, stored them through cool, dry winters and sowed them in the bare ground after the dangers of frost receded in spring. Far away to the south the inhabitants of the Andean foothills learned to cultivate tomatoes and quinoa from seed, and to grow potatoes from the tubers they kept from year to year.

Domestication was a haphazard process, advancing in one place, retreating in another, then remaining static for centuries as generation after generation followed the ways their parents had taught them. Even when domesticated forms of wheat, barley and rye were widespread and often predominant, it is still not unusual to find up to 50 per cent of the charred remains at Middle Eastern archaeological sites composed of weedy, unimproved forms. People continued to harvest seeds from the wild goat

1 The wild and uncultivated areas of the Golan Heights and the regions around the Sea of Galilee are places where seed collectors search for the wild relatives of familiar crop plants, such as wheat, barley and beans.

2,3 *Opposite* and *above* Einkorn, one of the earliest of the cultivated wheats, is found mixed with other wheat relatives at the margins of cultivated fields. The ears of einkorn are composed of numerous flowers (or spikelets) each of which produces a single grain. Compared to other cultivated wheat, einkorn ears are rather delicate. Thousands of years ago, einkorn was widely cultivated. Today, einkorn's cultivation is restricted to isolated regions in India and the Mediterranean.

4 *Left* Emmer wheats are larger and more robust than their einkorn relatives. Furthermore, they have four complete sets of chromosomes, rather than the two sets found in einkorn. Durum wheat, which is widely grown for the production of pasta and semolina, is a type of emmer.

5 *Right* Bread wheat accounts for 90 per cent of world wheat production. It evolved in cultivation from emmer wheat, and has six complete sets of chromosomes. Bread wheat incorporates genetic material from at least three different species of wheat.

6 Teosinte, the wild relative of maize, has grains arranged in many small spikes on branching stalks. In contrast, maize has a single main stalk and the grains arranged in a few, easily harvested cobs.

7 Cobs of modern maize differ greatly from race to race in shape, size and colour. This is the result of thousands of years of selection by humans.

grasses, wheats and barleys, probably because in many places, as they still do, they grew so thickly over such extensive areas they yielded a useful crop. We would call such plants weeds, but the concept would have meant nothing to people who had gathered them for millennia, and still did so as and when opportunities arose.

The first farmers, whether in the Middle East, Mexico or China, harvested seeds from an assortment of different plants depending on where they lived, but gradually domesticated forms became grown widely. Other crops, the 'hangers-on' known by archaeologists as secondary crop plants, were the source of many of the vegetables, herbs and annual flowers grown in gardens today. Anything edible was welcomed in the fields and used, as it still is today in many parts of the world. Some, such as fat hen, spinach, amaranth, Good King Henry and other leafy plants, including lettuce, were gathered while the cereal plants were still immature and eaten as green vegetables. Various onions, leeks, rocket, dill, fennel, chervil and herbs of one sort or another with strongly flavoured foliage added relish to the diet; radishes, carrots, salsify and other root crops were pulled as and when they were ready. Lentils, peas and many kinds of beans habitually grew among the cereals. The seeds they produced were harvested with the grain, threshed with it and contributed to the harvest as a matter of course, enriching the people's diet in the process.

The cereals themselves in the Middle East were a mixed lot. They derived from numerous wild species, sometimes including hybrids between wild forms. From time to time, when seeds were obtained by barter or other means from neighbouring groups, the mixing of genes from different local forms led to the appearance of new strains. Some would take their place among the cultivated plants; many would remain as 'hangers-on' or weeds, growing alongside and among them in the fields. Numerous cultivated Middle Eastern plants, including wheat (5), barley, rye, oats, carrots, radishes and lettuces, grow in association with weedy relatives. In years when pestilence, disease or drought decimated cereals, the greater resilience of some of these plants provided the means for a bare survival. Today we know that they also played crucially important roles in the evolution of crop plants, providing sources of resistance to new strains of disease and tolerance to drought, salt or frost.

Even at this early stage of plant domestication, the cultivated plants in the fields around communities had an impact on the neighbouring flora. Methods of cultivation were rudimentary; liberal views prevailed about what was sown and what was harvested. Little attention was paid to removing plants that were less than desirable, but the simple husbandry involved in tilling the soil, sowing, harvesting and storing seed tipped the balance against some species and favoured others. Many of the plants preferred by the gatherers failed to compete in cultivated fields. Their seeds did not germinate at the right times or their seedlings could not cope. Such plants became scarcer and scarcer until they were seen no more in the fields, and found only among the wildflowers in uncultivated areas.

Others easily made the transition from wildflower to cultivated plant and flourished. Each year their seeds were harvested with the grain and stored beyond the reach of insects and rodents, which would seek them out and devour them if they were lying unprotected in the soil. The seeds were sown at an appropriate time under conditions that gave them the best possible start in life, and when seedlings emerged they were protected from grazing animals when they were most vulnerable. Generation by generation, the plants responded to the farmers' support, losing as they did so some of the competitive edge which had enabled them to hold their own in the challenge of natural conditions. Released from the pressures of natural selection they changed, becoming less specific, sometimes more indulgent, less finely tuned to cope with the pressures and vicissitudes of independent existence. When, as was often the case, domesticated forms in the fields grew close to their wild relatives in the surrounding countryside, the wild and the cultivated forms interbred as bees, butterflies or the wind transferred pollen from one to the other. As a result, the less competent genetic types of the cultivated plants became incorporated in the wild populations, reducing their ability to survive. From time to time it is likely that some wildflower would be critically undermined by this influx, its resilience to adversity reduced, and become extinct. A grass called teosinte (6), which has played a very significant part in the evolution of cultivated maize (7), still grows around the margins of fields in the highlands of Mexico. Teosinte has large grains partially enclosed in husks as primitive kinds of maize may have been.

It has been a continued presence in and around maize fields throughout the history of maize cultivation, and there is no doubt that its genes are irretrievably part of the inheritance of every corn on the cob.

When human populations were small and widely scattered, the simplest methods were sufficient to gather what food was needed. People had few incentives to change – much of their diet was supplied by the game they hunted, and they could obtain the seeds they needed without much trouble. But whether people sought it or not, settlements in the Middle East, however rudimentary and temporary, led inadvertently and inevitably to change. As people became more dependent on gathered grains their numbers increased, and their more sedentary lifestyle put greater pressures on the countryside around their dwellings. Trees and bushes were cut for firewood, and the grasses moved in to occupy the spaces created. Sheep grazed the wildflowers, goats browsed the scrub, and year by year the grasslands around the settlements spread. People would prefer higher yielding, more easily threshed, more palatable kinds of cereals and, as they carried them back to their settlements, spilt grain concentrated these more favoured kinds close to where they lived. One of the preferred sorts was a hybrid, this time between the semi-domesticated, hybrid wheat known as emmer and another goat grass. The result of this benign interbreeding was the ancestor of the bread wheats we grow today.

More space for more productive kinds of grass, the advent of grass forms which yielded their grain with less effort and the construction of grain stores all made fields more productive. Settlements grew larger and more populous with better and more secure supplies of food, developing increasingly complex social systems to provide the security needed to hold and farm land. Eventually these changes led to the creation of the multifaceted urban communities we call cities. The decoy had closed. Henceforth the quality of our ancestors' lives and the survival of their societies would be inextricably bound – as ours still are today – to their ability to care for the seed and on their skills as husbandmen, bound to the soil.

Farming in the Middle East originated as an inadvertent consequence of the collection, storage and use of cereal grains for food: it owed little or nothing to deliberate intention. Nobody long, long ago picked a stalk of goat grass or looked at an ear of wild wheat and thought, 'What a useful kind of plant this could be if only we could make it bigger, better and more palatable.' We did not choose which plants to grow. Nature presented us with a list of what could be grown and said, 'Take it, and become farmers; or leave it, and continue to live off what you can hunt and gather'. This decisive point in human development occurred because the plants showed us what we could grow, which actions produced a crop and which failed to do so – not because early farmers mastered mysteries for themselves. They did not turn wild grasses of little promise into cereals capable of supporting civilizations. That was done for them by spontaneous hybridization, a process they could not understand and of which they were probably unaware. Had people kept an account of such things, they would have observed that, as dependence on cultivation increased, the number of plants that mattered became fewer and fewer. In the already distant past their ancestors had gathered scores of different plants. Yet eight thousand five hundred years ago a farmer in the Middle East grew perhaps a dozen different plants to provide the staples, and perhaps a score or so more as incidental food, for the entire community. Most of the plants that ancestors had gathered no longer played any part in their daily lives, because that was what the plants dictated – not because the farmers had decided they did not want them, or their families refused to eat them.

Most of these plants eliminated themselves because they did not possess qualities that made them amenable to cultivation. Those that did, including wheats, barleys and other primitive cereals, produced seeds which were easily handled (not too big and not too small), remained reliably viable for several years and germinated at the right time and in sufficient quantities to make cultivation a practical proposition. In addition, as the years went by, hybridization and selection introduced or reinforced qualities which made them more amenable, more useful and more valuable to farmers. These improvements went beyond obvious increases in grain size and cropping capacity, which resulted from combining large-seeded goat grasses with small-seeded wild wheats. They included anatomical changes by which the ripe grains were held more securely in the ears until they were threshed and, when threshed, separated readily from the husks. Physiological changes that led to increased frost hardiness or drought tolerance were also involved, as were metabolic changes which increased the proportion of protein, notably gluten, in the grains. This gave dough made from their flour the elasticity needed to make good bread.

Archaeologists have taken the trouble to work out the story of wheat because it happens to be the number one support of the human race across vast expanses of the globe's surface. However, what wheat was to become had little or nothing to do with its admission to the ranks of cultivated plants in the first place. Nobody foresaw its future; it created that for itself, partly because its wild progenitors were naturally endowed with qualities that made them amenable to cultivation in the first place, partly because these qualities were subsequently enhanced by a remarkable series of events involving hybridization and chromosome changes which mixed genes from different plant species. Last but not least, it happened to be a plant eminently worth cultivating. Similar stories of chance occurrence and spontaneous hybridization lie behind the domestication in the Middle East and elsewhere of many long established garden plants. Their ancestors combined qualities that made them useful with attributes that made it possible to cultivate them. It was all done so effectively that when deliberate attempts to breed improved plants started around two hundred years ago, we found we already had almost all the basic forms we needed to hand – not just of wheat but of hundreds of other plants grown on farms and in gardens. They had been domesticated inadvertently because once upon a time their ancestors had been part of the daily haul gathered from the wild.

The domestication of wheat in the Middle East was one part of a global transformation that started some twenty thousand years ago. Long before then humans had paddled, rafted or drifted across the Straits of Timor to Australia, and then moved on to Tasmania. Between twenty thousand and sixteen thousand years ago, small bands of Asian tribespeople crossed the barren wastes of grasses and sedges which then bridged the Bering Straits, and went on to colonize the Americas. Humans had occupied every quarter of the globe, apart from inaccessible islands such as Madagascar and New Zealand and the inhospitable wastes of the Arctic and Antarctic, and every-where pursued the lifestyle of hunters and gatherers. Then, over a period of some six thousand years – an age by historical standards, but the blink of an eye in the course of human existence – the inhabitants of widely scattered parts of the world discovered the power of seeds. In one region and another, and typically in various locations within each, the processes through which wheat, barley, lentils and rye became cultivated crops in the Middle East were repeated. In each place revelations of the powers of seeds opened the eyes of nomadic hunter gatherers to the secrets of preparing soil, sowing seed, cultivating, harvesting and storing crops.

In every one of these places people had lived for thousands of years by gathering what they could find from wild plants. Then, over a period of less than ten thousand years, people learned to cultivate the plants on which they depended instead of leaving everything to nature. How can one explain a zeitgeist so powerful and pervasive that it embraced people living so far apart? What wind of change could have transformed the lives of human beings in so brief a period?

Such questions may be imponderables, but a sea change did take place in these peoples' cultures and their attitude to the world about them. They had learned how to make the most of what nature provided, instead of making what they could from what nature offered. People had begun to manage the countryside in the immediate vicinity of their settlements to make it more suitable for the growth of the plants. They learned to care for the seed – storing it, sowing it, cultivating the seedlings and eventually harvesting the crop produced for their own ends. Some was consumed, but enough held back to produce a crop the following year. The mutually sustaining alliance between humans and plants, known as cultivation, changed both partners profoundly.

By taking responsibility for the care of the seed, the inhabitants of a few comparatively limited regions in widely separated parts of the world exchanged dependence on nature for more complex lifestyles. In a settled, farming society, prosperity and wellbeing depended on how successfully individuals organized themselves. An observer of events in the Middle East during this period would conclude that the inevitable consequence of settling down was the founding of cities. A contemporary observer in North America would see far less evidence of this and would be unlikely to draw such conclusions. The reasons why cities arose in one place but not in another are among the mysteries of the human condition.

Three crops were pre-eminent – wheat, maize and rice. Each was to become the staple food of people living across vast tracts of the world's surface, supplemented and complemented by a select number of other grains, as well as beans, squashes, potatoes and other root crops, which varied from place to place. Each appears to have been, if not the essential, at least the strongest predisposing factor leading to the development of farming and the establishment of settled communities. A cursory survey of the subsequent history and social development of the consumers of wheat, maize or rice reveals huge differences in their ways of life and particularly the extent to which they formed urban communities. Were these differences due to the geography and other physical conditions of the places where people lived? Or did the crops themselves create situations that defined the subsequent development of those who domesticated them?

Geography provides the seemingly obvious explanation, but is the variation between geographical conditions in different parts of the world sufficient to explain the magnitude of the differences between them? North America and Eurasia are both in temperate regions and share much in common climatically, topographically and geographically. Each has river valleys, deltas, harbours, plains, mountain pastures, steppes and forests. Conditions similar to those that nurtured the cities of Mesopotamia, the Indus Valley, even the delta of the Upper Nile can be found in North America. But, save the possible exception of what appear to have been mainly cultural centres, such as Cahokia at the junction of the Missouri and Mississippi rivers, cities or even large towns comparable in function or size to those of the Old World are notably absent. Examination of the physical features of the two land masses does not account for the extraordinarily different courses of development of their inhabitants.

Urban civilizations developed within a comparatively short time in a seemingly inevitable progression after the domestication of wheat – and its ancillaries barley, rye and pulses – across the whole extent of the Middle East. From the Mediterranean to the borders of northern China, the Indus Valley and North Africa, empires and hegemonies waxed and waned. Urban

27

civilizations also developed in tropical parts of Central and South America though, unlike those in the Middle East and the Indus Valley, cities there tended to be religious foundations and places devoted to ritual. Nevertheless, they were home to large numbers of artisans, tradesmen, administrators, labourers, workmen and porters – besides the priests and those concerned with religious rituals – who lived in conditions analogous to those associated with city dwellers elsewhere. The main crops in tropical regions included manioc, sweet potatoes, tomatoes, capsicums, avocados and cacao, in addition to maize, squashes and beans. These crops provided subsistence throughout the year, in contrast to the temperate agriculture of the Middle East and North America, which is seasonal with periods when little or nothing can be gathered from the fields. Consequently, food which can be stored throughout the off-season plays a much more significant part in the survival of people in temperate regions.

In parts of North America where tropical crops could not be grown, but where maize with squashes and beans could be grown during the summer, alternative, non-urban civilizations developed. These were totally different in character from those of Eurasia. When Columbus made his voyage to the New World, Europe had about twenty cities housing between fifty thousand and one hundred thousand citizens, some eighty more with between twenty thousand and fifty thousand, and umpteen more with populations of ten thousand to twenty thousand. At that time, across the entire extent of the region now known as the United States of America, there was not a single settlement with more than a few thousand inhabitants. Many settlements were graced with the title of town, but none displayed the complex social organizations or hierarchical forms of administration that distinguished such places in Europe or western Asia.

The indigenous peoples of North America, although they constructed no cities, developed organized and coherent alternative forms of civilization with powerful, cultural and artistic traditions. These structures extended beyond membership of tribal groups to affinity with long established allies across extensive tracts of land – not dissimilar to concepts of nationhood possessed by many countries of the Old World. In Eurasia such situations were catalysts for the foundation of cities as centres of trade, national identity, ports or military strongholds – but not in temperate North America. The inhabitants of many parts of the North American continent, especially east of the Mississippi river, often produced substantial quantities of maize, beans and squashes in fields around their settlements. They lived in village communities and supplemented the produce of their fields by hunting. Yet, apart from the Pueblo Indians in the southwest, settlements in North America were never more than semi-permanent. The materials used in the construction of the houses, and the ways in which they were built, result in few permanent remains to indicate the nature of dwellings in the Old World. Even east of the Mississippi river, where the houses were less ephemeral, they were still readily replaceable when a move to another location was necessary – the antithesis of the principles upon which towns and cities are constructed.

The different approaches to city building and urban civilization among the inhabitants of Eurasia and North America reveal positive attitudes to urbanization in the former and negative attitudes in the latter. Both views are so deeply rooted that they might almost be described as instinctive. The antiquity of the origins of the differences in outlook were suggested in 1906 by Pleasant Porter, the mixed-race chief of the Creeks, to a Committee of the US Senate. During proceedings leading to the declaration of statehood for Oklahoma, he analysed the situation as 'a complex problem, gentlemen… You are the evolution of thousands of years, and we the evolution of thousands of years, perhaps…We both probably started at the same point, but our paths diverged, and the influences to which we were subjected varied, and we see the result.' Could these influences have originated in the earliest days of agriculture as each group became familiar with, and accommodated itself to, the nature and demands of the staple crops which provided its daily bread? If so it begs the question: what qualities endowed wheat with the power to create cities that were not present in maize?

Domestication is a two-way process. Each crop makes particular demands and offers its own opportunities, methods of cultivation, levels of cooperation with neighbours and the ways in which it can be prepared as food. Social aspects affecting its consumption differ one from another, and influence the development of the society as a whole. Those who domesticated and depended for their daily sustenance on wheat, maize or rice responded to the crops they grew and the grains they ate, and the ways in which they responded endowed them with different expectations and ways of living.

Wheat is a sociable crop. In order to reap the rewards of growing wheat extensive areas have to be cultivated, sown and harvested – something that can only be accomplished through cooperation and effective social organization. Those who cultivated wheat, even in the most primitive way, would inevitably have had to create the foundations of the structural framework of governance. They would need to maintain law and order and to establish a form of social hierarchy – the precursors of successful urban development. In contrast, maize growers are not compelled to cooperate, though they may benefit from doing so. In many communities in North America, in fact, maize growing was a joint activity in which everyone combined to cultivate a common field, extending sometimes to hundreds of acres. Though attractive to crows and rats, maize can be, and very often is, successfully grown on a small scale. A peasant farmer, even on a very small holding, can reasonably expect to grow enough grain to feed his family, and provide a small surplus for sale or exchange.

Wheat, specifically the flour made from it, is an artisanal commodity. It provides scope for technique, know-how and imagination in ways which are not an option for those who depend on rice or maize products. In the hands of skilled craftsmen, flour is the basis of a remarkable range of appealing products, ranging across breads of many kinds, pastries, cakes, puddings, pies, pizzas and pastas. Rice and maize flour provide opportunities for a tiny fraction of the imagination and skill that can be expended on wheat flour. Village bakeries have been at the heart of village communities for centuries, providing the focal points of social interaction in towns and cities. Pastry cooks and others skilled in the ways of working with wheat flour are, and have long been, essential and valued in the entourages of rulers and people of influence. Hominy grits, succotash, polenta and mealie porridge not only bear no comparison in versatility and palatability to wheat products, but are also, wherever the last are available, frequently the food of the rural poor, urban labourers and slaves – the props, not the makers of societies.

From the earliest days of its cultivation, wheat encouraged cooperation and specialization. It fostered the development of a social organization in which farmers grew the grain, millers ground the flour and bakers baked – releasing the energies of other members of the community and enabling them to play specialist roles. Wheat, maize and rice all provided the foundations upon which settled communities could be built, but whereas the foundations provided by maize and rice were sufficient to build walls, wheat's additional qualities provided the keystones of arches to support the edifices of urban civilizations.

For thousands of years, until well into the twentieth century, peasant farmers throughout the world harvested their grain from whatever combination of desirable plants happened to produce it. The seed that they sowed for the following crop would be a sample from whatever they had gathered previously. In the Middle East – and across Europe and many parts of North Africa, as agriculture spread – wheats predominated in some locations, barleys in others, rye in cooler, drier parts and oats in places where short growing seasons and wet summers provided less favourable conditions for wheat or barley. Occasionally a farmer would introduce seeds bought in the market or exchanged with neighbours, but in general they relied on self-sufficiency, putting their faith in whatever had survived the best of years on their holdings. This was the prehistoric model, and it was the crucible that created the farm crops and vegetables we grow today.

Wheat, barley, maize, rice, rye, millets and sorghum (the staple grain crops of the world), soybean, peas, lentils and beans of many kinds, sweet potatoes, potatoes, squashes, dates and bananas all come to us with the compliments of our Neolithic ancestors. Almost the entire spectrum of cultivated fruit, vegetables, cereals and other crops originated long before science played any part at all, and indeed long before we had the slightest understanding of how seeds worked or the potential they concealed.

Hunter gatherers learned to gather seeds and sow them when the time was right to produce another generation of the same mixture of plants. Gardeners do much the same today, though most of us buy the seeds we sow rather than collect our own. Our ancestors regarded seeds as dehydrated plants, or rather convenient kinds of buds, and were satisfied when they came up and fulfilled their expectations by producing plants like those from which they had been gathered. Most of us are happy today when the seeds we sow germinate and the seedlings grow into plants that look something like the pictures on packets in which they came. Although our ancestors ten thousand years ago did not understand the biology of germination and the complexities of plant sex, they may have understood the subtle needs of seeds more than many of us today.

CHAPTER TWO

The Genie Released

What do you see when you look at a seed? Neolithic people would have answered, 'Food', with nothing more to be said. The first farmers learned to use them as conveniently packaged dehydrated plants, the seed corn of the crops on which societies were built and, in many places, urban civilizations constructed. Recently we have taken them apart, penetrated their innermost secrets and unravelled the meaning of the genes arranged along the DNA strand and packed into the chromosomes of the their cells' nuclei. When we look at seeds today we see the means by which plants express their genes. This is the knowledge that enables us to exploit them.

In the last few decades we have gone further, breaking the bonds that bound us to nature's limitations to explore a world with no such restraints. In the twenty-first century we can transfer genes between different, even totally unrelated, organisms, offering seemingly limitless prospects for further benefits. Yet such new horizons also pose a danger – the risk of inflicting previously unimaginable damage on the world around us. Do we have the foresight, wisdom and restraint to reap the benefits of our new-found ability to transfer genes between organisms? In doing so, can we avoid the disasters forecast by those opposing what they regard as unnatural and forbidden genetic modifications?

The answer to that question lies in the future. However, in order to discover how we have reached this position, we need first to trace the steps taken by the gardeners, farmers and botanists who recognized that the enigmatic exteriors of seeds masked unsuspected mysteries. These objects were to prove more interesting than anyone could ever have imagined.

Two hundred years after the European Renaissance, and one hundred years after the so-called rebirth of science, biological sciences remained enthralled by bonds woven by ancient Greek philosophers. Professors of natural philosophy at universities across Western Europe saw nature through a lens that reduced their observations to abstract philosophy. Rather than draw conclusions from what they observed, they were taught to deduce what went on in the world of plants and animals according to the precepts of Aristotle and his disciple Theophrastus, overlaid by theological dogmas based on biblical accounts of the Creation. The result of this blend of philosophy and theology was a view of the world far removed from reality.

Sexual reproduction, the philosophers had ordained, was the prerogative of creatures capable of movement – a conclusion based on the apparently reasonable supposition that sex required the ability to move around in search of a mate. Plants were immobile, with no means of finding mates, and must, inevitably, be asexual. Never mind that numerous plants, including such conspicuous and familiar trees as the date palm, exist in two forms. One was popularly referred to as male, the other as female, and without both, as every owner of a grove of dates knew, the trees were barren. Never mind that gardeners recognized the need for pollination, though they might be a little bit hazy about, and probably not much interested in, just what form the sex lives of their vegetables took. For two thousand years men of learning and academics closed their eyes to what they saw every time they looked at a flower. Somehow they managed to disregard errant ideas that the very visible organs that make up a flower are blatantly sexual in purpose. Sex in plants was a figure of speech and nothing more; anything else was the stuff of dreams and the product of overheated imaginations. While these attitudes prevailed botany remained a sterile science, with few practical applications apart from ancient ideas about the roles of plants in medicine. Reluctantly and by slow degrees, through almost all the eighteenth century and much of the early nineteenth, the genius of a few free-thinking, eccentric individuals brought guardians of the natural sciences eventually to appreciate that truth lay in what they could see, not in the writings of long-dead men from another time.

In the final decade of the seventeenth century Rudolph Camerarius, professor of natural philosophy in the University of Tübingen and director of the Botanic Garden, investigated a belief held by many gardeners – that pollen played an essential part in the production of seeds. Instead of philosophizing, he looked around for a practical way to test the possibility, and chose for his experiments flowers of plants which, with the benefit of hindsight, seem remarkably unsuitable for such an experiment: annual mercury and mulberries. Camerarius removed the anthers from some flowers and they produced no seeds; other flowers to which he applied pollen did produce seeds. Encouraged, Camerarius repeated his experiments on flowers of spinach, castor oil and maize, concluding that pollen was indeed essential for seed production. Anthers in which pollen is produced could only be male sexual organs. Ovaries and the styles, which bear the receptive stigma, to which he had applied pollen, must play the female part.

Camerarius had demonstrated that the male and female roles attributed to anthers and styles were not merely figures of speech. It was a break-through. He decided that pollen must be a kind of proto-embryo, in line with the ideas of the 'spermists' (who believed that during sexual intercourse a man contributed a minutely formed but complete infant – a homunculus – while the woman merely provided a safe, warm and nourishing womb where the baby could develop). The styles of flowers, he concluded, were probably hollow tubes down which pollen grains found their way into the ovaries; there they were nourished during their development into seeds. A simpler explanation, and one closer to the truth, was that the ovules themselves contained immature embryos and pollen played some such role as exciting them into action and eventual development into seeds. In a less androcentric world this might have been more sympathetically considered.

Camerarius's observations were largely ignored after their publication in 1694, the usual supposition being that he published them in obscure journals where they were overlooked. Yet Rudolph, his father before him and his son after him were successively professors at Tübingen University; as notable members of an academic dynasty, their activities would not have passed unnoticed. If his peers failed to read, or understand, his reports, it could only have been because their academic preconceptions dismissed such observations as irrelevant. The philosophical debates of his fellow professors in the universities were blind to the absurdity of attempting to reconcile the obvious, the visible presence of male and female organs in plants, with the fabulous – the belief that plants did not possess sexual organs.

Such discontinuities between what is to be seen and what is to be believed call for mental gymnastics similiar to those with unquestioning belief in the Creation myth or intelligent design. Almost thirty years after Camerarius's demonstration that the anthers and styles of plants were quite evidently sexual, in 1723 a well-informed and cultivated academic, the philosopher Christian Wolff of Magdeburg, described in *Vernünftige Gedanken von der Wirkungen der Natur* ('Rational thoughts on the operations of nature') events leading up to the formation of seeds in the following way:

'We find inside the flower a number of stalks disposed in a circle and something at the top of each [the stamens and anthers] which is full of dust and lets the dust fall on the upper part of that which holds the seed; this organ is compared by some to the genitals of the animal, and the dust to the male seed; they think also that the seed is made fruitful by the dust, and therefore the embryo must be conveyed by the dust into the seed-case and there be formed into seeds.'

But then he added:

'Since all that has been hitherto adduced is found also in flowers which spring from bulbs and it is also certain that the leaves of bulbs have consequently embryos in them…it is easy to see that the embryos must come from the leaves of the bulbs. And since they could as easily be conveyed from there into the seed-grains with the sap as into the dust which is produced in the upper parts of the flower, I am inclined to think that this is the true account of the matter.'

Wolff goes on to speculate about how the embryos might get into the sap and concludes that 'it is certainly more credible that the embryos already exist in little in the sap and the plant before they are brought by some change into the condition in which they are met with in the seed and in buds'.

The next question he asks is where the embryos might have been previously: 'They must either lie one in another in a minute form…or they are brought from the air and earth with the nourishing sap into the plant.'

8 Insects are the most important plant pollinators. In this case, honey bees are visiting a member of the dandelion family. Masses of orange pollen are clearly seen filling the pollen baskets on the hind legs of the bee coming into land.

9 The catkins of hazel are adapted for the dispersal of pollen by the wind. Plumes of pollen are released from the mature catkins by the lightest of breezes.

Plate V

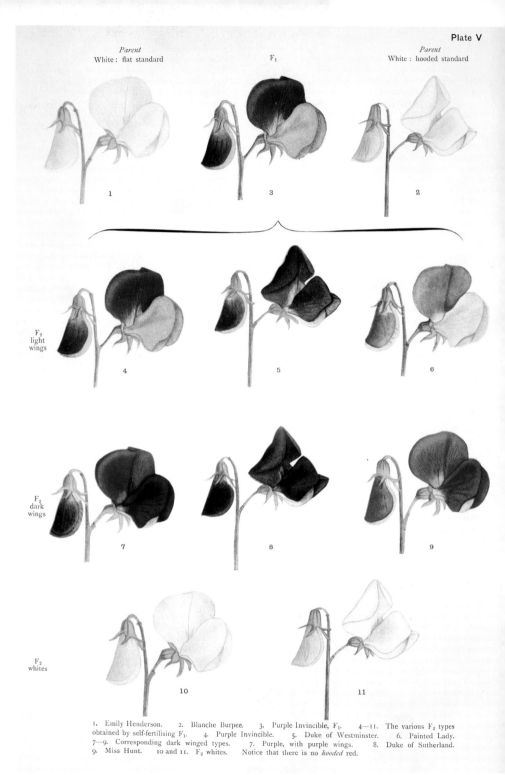

Parent
White : flat standard

F₁

Parent
White : hooded standard

1

3

2

F₂
light
wings

4

5

6

F₂
dark
wings

7

8

9

F₂
whites

10

11

1. Emily Henderson.   2. Blanche Burpee.   3. Purple Invincible, F₁.   4—11. The various F₂ types obtained by self-fertilising F₁.   4. Purple Invincible.   5. Duke of Westminster.   6. Painted Lady.
7—9. Corresponding dark winged types.   7. Purple, with purple wings.   8. Duke of Sutherland.
9. Miss Hunt.   10 and 11. F₂ whites.   Notice that there is no *hooded* red.

10 Artificially crossing plants, in this case peas, with different characteristics produces hybrids that can be used to deduce how particular characters are inherited, and uncover variation that was masked in the parental plants. Once the inheritance of a character is understood it is possible to breed plants with particular combinations of characters.

11 Gregor Johann Mendel (1822–84), the father of genetics, made controlled crosses between peas with different characteristics. By counting the numbers of the different types of offspring that were produced, Mendel deduced the basic laws of inheritance that bear his name.

12 Sweet peas were introduced into cultivation at the end of the seventeenth century from wild plants collected in Sicily. These first introductions appear to have had small, maroon or purple flowers. However, careful crossing and selection by gardeners, even without an understanding of genetics, has produced a vast array of different cultivars.

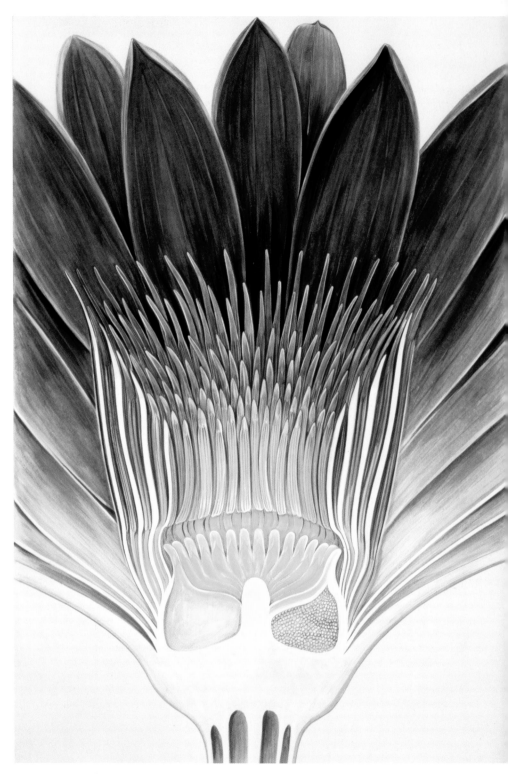

13 Vertical section through the flower of a water lily, painted by Arthur Harry Church. The gradual transition from protective sepals on the outside of the flower through attractive petals to the male anthers is shown. At the centre of the flower are the female carpels, which contain the ovaries, inside which are the ovules. Once the ovules are fertilized they will form seeds.

Even Wolff's elastic philosophical resources failed to contemplate the concept of a seed as a kind of Russian doll containing an infinite series of minute precursors of its descendants for evermore. He concluded that the more plausible alternative was to suppose that seeds were carried into the plant with the sap from the earth and air. In doing so he plumped for the wrong alternative. The infinite series, in a form he could not possibly conceive – a self-replicating double helix – was the answer to his conundrum.

While discussions about plant sexuality in universities continued to revolve around concepts such as those floated by Wolff for many years, gardeners had a more practical view of the roles of male and female organs of flowers on the plants from which they made their living. In the 1776 edition of the book he published with the support of other gardeners, *Every Man His Own Gardener*, John Abercrombie was in no doubt that although some gardeners considered 'male' flowers to be unproductive and removed them, they were in fact designed to impregnate the female flowers. For as, he wrote, 'In the early culture of cucumbers, etc, it is eligible to carry some of the males to the female flowers, previously pulling off the petal or flower-leaf of the male, then with the remaining anther or central part touch the stigma of the female, so as some of the farina or powder of the anther adheres to the stigma, a little of which being sufficient to effect the impregnation.'

This description of pollinating cucumber flowers is entirely comprehensible to anyone who grows melons, but because the cucumbers we buy have been produced without benefit of sex, it needs a little explanation today. In fact any cucumbers that are pollinated grow grossly pregnant with masses of inedible seeds and are ruined – hence the popularity of varieties that produce only female flowers, excluding any possibility of unwanted pregnancies. Abercrombie was referring to plants akin to the ridge cucumbers grown out of doors today and used for pickling. These, like their cousins courgettes and melons, develop fruits only after pollination.

Orthodox botanists at that time were largely preoccupied with devising classifications which arranged plants in logically satisfying ways, and had little interest in how they worked. The most successful classification was devised by Carolus Linnaeus, who introduced a sense of order into our view of the plant world based on the number and arrangement of the sexual organs of flowers. Linnaeus summarized much of the early work on plant sexuality in the 1730s, but he was primarily interested in the number and disposition of the anthers and styles so he could group plants in an orderly fashion. Linnaeus believed and argued that since the anthers and stigma were closer to a plant's 'sexual essence' than structures such as flowers or fruits, concentration on these characteristics should produce the 'best' plant classifications.

In the middle of the eighteenth century, botanists had no concept of the links between sex and inheritance, and since they considered seeds as more or less analogous to buds, there appeared little point in pursuing details. One exception was Joseph Koelreuter, professor of natural history at the University of Karlsruhe. He had studied at Tübingen some years after Camerarius had completed his experiments on pollination, and the connection may have stimulated his interest in pollen.

Camerarius had shown that seeds were produced only after flowers had been pollinated. The next step was to find out what pollen did, and to do that it was necessary to find out what part their sexual partners, the ovules, played. Koelreuter set out to discover whether some characters were inherited from the pollen, others from the ovules. This was a strikingly original question to ask during the 1760s. Over and beyond that Koelreuter's experimental approach – based on postulating a hypothesis then setting out to test it – was anathema to academics convinced that logic based on philosophical precepts, rather than experiments, was the way to discover truth. But did Koelreuter's inspiration spring from a heretical attitude to botanical teaching, or from his observations of the activities of gardeners in the university botanic garden?

Gardeners were becoming increasingly interested, though still tentatively, in the possibilities of producing improved varieties of fruit trees, vegetables and other plants by hybridizing one kind of plant with another. To them, the idea that certain characters were inherited from pollen and others from the ovules would have seemed an interesting, plausible and exciting possibility. In 1717 Thomas Fairchild, a nurseryman in Hoxton on the outskirts of London, had achieved fame – even a little notoriety – by crossing a carnation with a sweet william to produce a hybrid pink known as

'Fairchild's Mule'. Fairchild, a devout man, had qualms about what he had done, which he felt smacked of assuming the Creator's role. To assuage his conscience – and perhaps ease his way into the hereafter – he bequeathed £25 in his will to the Parish of Hoxton to provide an annual stipend of £1 to pay for a sermon on 'the wonderful works of God in the Creation'.

Before he could test his hypothesis, Koelreuter needed to find plants to act as male and female parents. They had to be sufficiently different for him to be able to recognize features inherited from one or other in the offspring – for example a green-leaved cabbage crossed with one with red leaves, or a cross between a larkspur with blue flowers and one with white. He went a step further, probably inadvertently rather than deliberately, and his first experiments were made with two species of tobacco. They duly produced seeds, but the plants he grew from them turned out to be sterile, rather as mules, the offspring of horses and donkeys (and indeed 'Fairchild's mule'), cannot reproduce. Subsequently Koelreuter extended his experiments to include garden pinks, stocks, henbanes and mulleins, among other flowers, with more success. Their seedlings, he observed, inherited the qualities of both parents, proving that pollen and ovule both contributed to the formation of seeds. He also noticed that he obtained seeds only when he crossed two closely related species. Mulleins pollinated by pinks were barren, but pinks crossed with other pinks or mulleins with other mulleins were much more likely to produce capsules filled with fertile seeds.

Koelreuter followed up his observations with further questions. How many pollen grains did it take to produce a seed? What went on after pollen grains were deposited on the stigmas of the flowers? The answer to the first question was that, although pollen appeared to be produced in large quantities, very few grains were sufficient to produce a seed. He managed to narrow the figures down to something like three seeds produced from five pollen grains, but never quite succeeded in showing that only one was needed, as is in fact the case. The second question proved more intractable, but what he saw convinced him that the entire pollen grain did not travel down the style – as the spermists had proposed.

The shortcomings of contemporary microscopes limited Koelreuter's explorations, but whenever he was able to observe what was happening he could be relied on to draw cogent and original conclusions. He defined the functions of the anthers, stigmas, styles and other structures found in flowers, including the nectaries. Before Koelreuter, nectaries had been regarded as strange, ambiguous structures involved in waste disposal. Koelreuter pointed out not only that they secreted nectar, but also that the nectar attracted insects, including bees that used it to make honey. Today we know that nectaries are diverse structures that secrete complex mixtures of sugars and other nutrients that act as a reward for pollinators. Furthermore, Koelreuter proposed that insects played a vital part in the sexual reproduction of plants and the formation of seeds by transmitting pollen from one flower to another (8). Those, too, were original ideas, dispelling beliefs held by some that insects simply went around robbing flowers of their pollen.

Mistletoe also attracted Koelreuter's curiosity, perhaps because of its oddity, but also for the more cogent reason that mistletoe plants are either male or female. He was able to show that female plants produced berries only when their flowers were pollinated from a neighbouring male. He also observed insects flying from plant to plant and acting as pollinators. Later he noticed something else – thrushes ate the mistletoe berries, then smeared the sticky substance surrounding the seeds on to the branches of neighbouring trees, where seeds germinated to produce seedling mistletoe plants. Such symbiotic interactions between plants and animals were the stuff of myths or unknown academically when Koelreuter published descriptions of these observations. Almost a century was to pass before Charles Darwin's observations made such interactions common knowledge.

All these discoveries excited little interest among his peers and, to Koelreuter's intense disappointment, lay gathering dust for many years. Academics in the universities refused to consider them on philosophical grounds, or at best accepted them with a grudging lack of understanding of their significance. Their refusals were reinforced by their commitment to the biblical account of Creation which condemned any suggestion of the 'creation' of hybrids as a heretical rejection of the concept of the constancy of species. This proved a forewarning of the furore later to greet the publication of Darwin's theories on natural selection.

Joseph Koelreuter had broken through the dogma that had restricted academic perceptions of the biological sciences for centuries, but his peers were still not ready to face the world without blinkers. Julius von Sachs, then chair of botany at the University of Wurzburg, pinpointed the situation vividly in 1875 when he wrote in his *Geschichte der Botanik* (*History of Botany*): 'On reading the observations of Linnaeus, Gleichen and Wolff on the sexual theory we step into a world of thought which has long been strange and is scarcely intelligible to us, and which in the present day possesses only a historical interest. Koelreuter's works on the contrary seem to belong to our own time.'

Outside the universities, men of a scientific frame of mind, practical farmers and gardeners were doing things equally strange and unintelligible to the academic mind. They were reaching conclusions wholly consistent with Koelreuter's discoveries, even though many of them remained blissfully ignorant of both the man and his work.

The men who would have brought balm to Koelreuter's wounded feelings included Christian Konrad Sprengel and Thomas Andrew Knight. The former was a botanist with a quirky personality and strikingly original mind; he largely resolved remaining questions about sexuality in plants, and the whys and wherefores of the astonishing diversity of flowers. The latter was a landowner, practical farmer, gardener and dilettante scientist who opened the way to using such knowledge in the pursuit of improved methods of farming and gardening.

Konrad Sprengel was a man apparently incapable of following rules or living an ordered life. After ordination as a priest he was entrusted with the pastoral care of the citizens of Spandau in Berlin. Fellow clergymen across the length and breadth of Europe, most famously his contemporary Gilbert White, curate of Selborne in southern England, indulged their interests in natural history, geology, archaeology and almost anything else which took their fancy, while still finding time to attend to their parishioners' needs. However, Sprengel's obsessive involvement with the earth-bound study of botany left him so little time for the spiritual needs of his flock that he failed even to deliver the customary Sunday sermons. His superiors sacked him and, deprived of his living, he devoted himself to his experiments on the reproduction of plants. He kept body and soul together by a little teaching, and by leading weekly botanical rambles into the countryside around Berlin – passing round the hat for a *groschen* or two from those who accompanied him.

Sprengel's obscure and tenuous lifestyle removed him from the society and influence of the botanical establishment, but it released powers of original conception that his peers failed to comprehend. The descriptions he wrote of his observations struck them as so surprising and conformed so inadequately with the philosophies of the ancients that they quietly and thankfully ignored them. There they lay almost, but not quite, forgotten until, fifty years later, Robert Brown, the most eminent, obsessively pedantic, British botanist of the time, and one of the few who had taken the trouble to explore such obscure corners, advised Darwin to read them when he was studying the extraordinary ways in which orchid flowers are adapted to attract pollinating insects. As Darwin commented later, 'The merits of poor old Sprengel, so long overlooked, are now fully recognised many years after his death.' Every dog has his day but it was Sprengel's bad luck that his was to come after his own demise.

Sprengel confirmed Koelreuter's demonstration that gene exchange (though neither had any concept of the existence of genes) was an entirely normal phenomenon. In fact, he chose the easier route of hybridizing different varieties of vegetables and other domesticated plants, rather than working with species. In consequence, although his observations may not have been quite so significant from a botanical point of view, they had much more practical potential. They opened the way for plant breeders to cross one form of lettuce with another, one kind of wheat with another, one apple with another, creating melting pots of desirable characters. These, if only they could be persuaded to come together in favourable ways to produce new varieties of lettuces, wheat or apples, would give rise to some truly remarkable vegetables. It was not as easy as that, as Sprengel's successors would discover, but the possibilities of selectively breeding new varieties of fruit, vegetables and crop plants had been revealed, with all its potential for good and ill. Sprengel also established the hitherto unsuspected – or at least unproven – fact that cross-pollination was the rule in plants, encapsulated in his comment that nature 'appears not to have intended that any flower should be fertilised by its own pollen'.

He went on to draw attention to the ways the colours of flowers, their forms and markings could be correlated with their abilities to attract the insects they relied on for pollination. Bees gather nectar and pollen for themselves and, in the process, pollen is transferred from the anthers to the stigmas of flowers. Most often pollen is transferred between different flowers, usually on different plants, and the plants are cross-pollinated. Occasionally pollen is transferred between flowers on the same plant, or within the same flower, and these plants are self-pollinated. The familiarity of this process and its function makes it hard to appreciate what a revelation it was at the time. Sprengel was the first to draw attention to the alliances between flowering plants and insects that have underpinned the evolution of both groups.

He also pointed out that not all flowers are colourful, and that those of grasses and many trees and shrubs are small, often to the point of insignificance. They produce no nectar, possess no brightly coloured petals or honey guides to point the way to visiting pollinators, and quite evidently do not attract insects of any kind. The pollen they produce is different too from that of more obviously floriferous flowers. It is produced in enormous quantities and is dust-like in its capacity to float in the air. Wind, Sprengel concluded, also played the role of pollinator, conclusively establishing firm links between form and function in plants (9).

As a result of literally thousands of observations, Sprengel demonstrated that the multiplicity of colours, shapes and curious devices represented in flowers were not simply expressions of the Creator's capacity for imaginative innovation, but responses to selective pressures. The corollary that flowers and insects had evolved together in intimate partnerships essential for the survival of both was an imaginative leap which Sprengel may or may not have made. We do not know. He was so disregarded in his time, so discouraged by the less than enthusiastic reception of his book, *Das entdeckte Geheimnis der Natur im Bau und in der Befruchtung der Blumen* ('The newly revealed secret of nature in the structure and fertilization of flowers') (1793), he gave up the idea of publishing a sequel and consoled himself with the study of languages. He died in 1816, largely forgotten by the academic establishment. His achievements would most probably have been forgotten with him if one of his former students had not kept his memory alive by publishing a description of his life and work three years later in the widely read journal *Flora*.

Sprengel's circumstances meant that although we are aware of the facts revealed by his observations, we have little idea what thoughts they may have raised in the mind of their discoverer. Was he content to regard the evidence he had revealed between form and function in flowers and the role of insects in their pollination simply as further proof of the ingenuity and attention to detail displayed by the Creator who originally furnished the world with an immutable set of plant and animal species – as the Church decreed? Or did Sprengel contemplate heretical thoughts that such links might be explained by selective pressures similar to those proposed by Charles Darwin some seventy years later? If he had such thoughts, like Darwin, he may well have been daunted by the prospect of challenging the established view of the primacy of Creation and the immutability of species. Darwin was held in deep respect by his peers and enjoyed the support of staunch friends in high places. Sprengel, regarded as an eccentric oddity, devoid of close associates and living some seventy years earlier in an even more strongly entrenched ecclesiastical society, would have been overwhelmed by the ridicule and censure such an outrageous assault on established thought would have provoked. He may well have thought it wiser to keep such beliefs to himself.

Koelreuter and Sprengel appear today to have left little room for further discussion about the sexuality of plants, and the parts played by anthers, stigmas, styles and ovules in the production of seeds. But there were still men in high positions in the scientific establishment who, despite the evidence of their eyes, clung to the philosophical argument that plants did not move and immobile organisms were sexless, ergo sex played no part in the life of flowers. Eventually, in 1830, the Academy of Sciences at Haarlem University offered to award a prize to anyone able to prove that sexuality did indeed exist in plants. Typically, the pragmatic Dutch made the prize conditional on the work leading to improvements in understanding which could be used to breed new and improved varieties of fruits and vegetable.

Seven years later the prize was awarded to the German botanist Karl Friedrich von Gärtner. Twelve years after that he published a vast monograph describing experiments involving nearly ten thousand attempts over twenty-five years to hybridize plants within and between more than seven hundred different species. Von Gärtner's impeccable academic credentials, as well as his unassailable diligence, finally convinced all but the most adamant adherents of the philosophical traditions of Aristotle and Theophrastus that sexuality was indeed a potent presence in plants. Pollination was almost invariably the prelude to the development of a seed. Hybridization between different species was possible, though limited in the extent to which it could occur. Offspring inherited qualities from both parents. By the middle of the nineteenth century academia had at long last been forced to recognize the seed's essential function as the conveyor of the plant's inheritance.

Having discovered how seeds are produced, scientists were now ready to find out what seeds did. They had set out believing seeds, like buds or runners, were just another means of reproducing their parents. But the offspring of buds and runners grow up in the precise image of the plant that bears them. Seeds, it transpired, are an amalgam of two plants – the plant that bears them and the plant that produces the pollen, which fertilizes the ovules from which the seeds develop. If they come up in the precise image of the plant on which they are produced, it is only because, as is usually the case with a natural species, both parents are so similar that the contributions of one are indistinguishable from the inheritance of the other. When the parents are dissimilar – interfertile species, perhaps, or different varieties of a vegetable – their offspring inherit characteristics of both parents.

This was science with a practical application, far removed from the philosophical carousels of academic botanists. It called for men – for many years to come woman played little part in plant breeding (women did not become significant, practical plant breeders until the early twentieth century) – who combined the practical approach of farmers or gardeners with the intellectual curiosity of scientists. The man who answered the call, without waiting for Karl von Gärtner to persuade the prize-givers at Haarlem University of the practical possibilities of hybridizing plants, was almost the contemporary of Konrad Sprengel, but different in character and circumstances in almost every way. Thomas Andrew Knight was a Herefordshire country gentleman. He had inherited substantial independent means, occupied a respected position in the county where he lived, and would become the close personal friend of influential people in positions of great authority in London.

Knight's grandfather had run an iron foundry on the banks of the River Severn in the early days of the industrial revolution, and used the fortune he made to set his family up as landed gentry. After going to school first in Ludlow, then in Chiswick in west London, and finally graduating from Balliol College in Oxford, Thomas Andrew returned to his roots in the neighbourhood of Ludlow. Here he married and settled down to live the life bequeathed him by his grandfather on an estate at Elton. He was a man of modest ambitions, rather diffident with a quiet, gentle temperament, ideally suited to the life of a gentleman farmer. Portraits reveal an amiable-looking man with a reassuringly solid demeanour – someone to be depended on not to let you down, but less obviously likely to display great originality of thought or imaginative insight. Yet Knight's solid outside appearance masked a most active, enquiring and speculative brain. Not content to watch his crops grow, harvest his wheat and cabbages, pick the pears and apples in his orchards and make cider, he could not see a plant without asking questions about it. Why did roots grow down and shoots up? How did the sap move around the plant? Why were stems and leaves shaped as they were, and why did they vary so much one from another? When he harvested his apples, pears and strawberries, his mind turned to the possibility of producing better varieties with larger, sweeter fruits that cropped more heavily or more reliably.

Knight set about finding answers to the questions by using the plants growing in his fields and glasshouses to study the effects of nutrition, methods of limiting losses from pests and diseases, ways and means of forcing vegetables, and other directly practical aspects of plant growth. He delved deeper in more scientifically orientated experiments designed to investigate physiological aspects of plant growth and the ways in which plants respond to gravity or electrical stimuli. Knight's fusion of traditional gardening practices with botanical studies on pollination, sexuality and inheritance marked the birth of the science of horticulture. Most relevantly in the context of this book, he set about producing improved varieties of apples, pears, cherries, plums and various vegetables by making crosses between promising parents.

Knight's most notable successes arose from one of the first attempts to combine the very different qualities of the North American common strawberry and the Chiloe or beach strawberry. The former was one of the first plants to attract the attention of the settlers at Jamestown in April 1607 – on the first day of the first permanent English settlement in North America. Master George Percy recorded his impression of Virginia as a place of 'fair meadows and goodly tall trees…we passed through excellent ground full of flowers of divers kinds and colours…Going a little further, we came into a little plot full of fine and beautiful strawberries, foure times bigger and better than ours in England.' The other ancestor of our cultivated strawberries – the beach strawberry from the Pacific shores of the Americas – produces large, dense fruits, which resist rough handling, but are dry and almost tasteless. Strawberry breeders today still seek the Holy Grail represented by the perfect combination of the tasty succulence of the Virginian strawberry with the travelling qualities of the beach strawberry.

Everyone who sets out to breed new and better fruits or vegetables encounters the discouraging fact that an awful lot of frogs have to be kissed and discarded before they find Prince Charming. The vast majority of seedlings fall far short of, or barely compare with, their parents, and even when a seedling does produce fruits with exceptional flavour it seems perversely and invariably to be cursed with intolerable disabilities – susceptibility to disease, infertility or worse – which make it useless.

Knight's two strawberries, the Downton and Elton, were two princes among four hundred frogs. Two hits from four hundred would be regarded as a phenomenal success rate by a strawberry breeder these days. Nevertheless, Knight found the odds he faced long enough to make him think about ways to reduce them. Perhaps he would do better if he knew more about the rules, if any, which governed inheritance. He looked around for a plant which seemed more suited to serve as a guinea pig for his experiments than strawberries, and came up with the idea of using garden peas.

Peas were an inspired choice. Long after Knight was dead, they would enable a man in a distant country to discover precisely what Knight had hoped for from them (p. 57). Pea flowers are almost invariably self-pollinated, and generations of inbreeding have produced uniform strains in which every plant has the same genetic make-up. They flower and produce seeds within weeks of being sown. There are numerous varieties with distinctively different characteristics: some grow shoulder-height, while others barely brush the knees. Some produce green peas, others grey. They can be smooth or they can be wrinkled, and while most varieties have white flowers, in a few they can be purple. As a bonus the flowers are easily manipulated in the course of a strict regime of controlled pollinations, and there are only half a dozen or so peas in each pod. This provided a manageable number, in contrast to the hundreds of seeds per capsule that Knight would have had to cope with if, for example, he had tried to work with poppies.

He made his first crosses during 1787, sowing the seeds they produced the following year. He recorded how the plants that grew from them resembled or differed from their parents. The first thing he observed was that some qualities appeared to be lost, others perpetuated and even accentuated. For example, when he crossed a dwarf with a tall pea every seedling grew up tall. Not only that, they outgrew even their tall parent. When he crossed a variety with wrinkled seeds with one whose seeds were smooth, every seedling produced pods filled with smooth seeds. The following year he sowed the seeds produced by the hybrids and, lo and behold, qualities that seemed to have been lost reappeared. There were dwarf peas as well as tall ones; pods filled with wrinkled seeds as well as pods with smooth ones.

In Knight's description of his experiments (entitled 'An Account of the Fecundation of some Vegetables'), published in 1799, he was not able to explain the behaviour of the peas. But he had stumbled over the foundation stones of the science of genetics, long before either the science or the concept of genes had been conceived. He had established that different characters are inherited from one generation to the next, cast some light on the patterns of inheritance and encountered the phenomenon of hybrid vigour. He had revealed – without actually discovering – dominant and recessive characters, and demonstrated that particular characters are inherited as distinct features. Years later the way the integrity of different characters is maintained would be known as segregation.

Whatever he had hoped for, Knight's observations on peas would never have been much help in his apple and strawberry breeding projects. This is because apples and strawberries have more complex inheritance patterns than the unusually straightforward systems of peas. Such projects had, however, confirmed his belief that different characters are passed on from generation to generation, and justified the good sense behind breeding from parents possessing the qualities he hoped to obtain in his seedlings. It was still too early, and understanding of the principles involved was too sketchy, for scientific principles to be applied to plant breeding, and Knight continued to be guided entirely by his instincts in his choice of parents. In doing so he acted no differently from many plant breeders, even today.

Fortunately, Knight published accounts of what he was up to, otherwise this remarkable man might have continued to pursue his experiments and his plant breeding programmes almost unnoticed beyond the borders of Herefordshire and Shropshire. As it was, his activities attracted the attention of Joseph Banks, a man of such enormous energy that it sometimes seems he single-handedly stirred the world of gardening throughout the British Isles into a ferment that has never subsided. Banks, then president of the Royal Society and the *de facto* director, on behalf of the king, of the Royal Botanic Gardens at Kew, recognized the originality and significance of Knight's experiments and observations. He persuaded him to come to London in 1796, where he presented a paper to the society on 'The Grafting of Fruit Trees'. Banks's enthusiastic and forceful personality was just the sort of spur needed to overcome Knight's diffidence and give him the confidence to embark on a substantial role in the country's horticultural affairs. Such a position could only have been buttressed in 1805 by the decision of his older brother, Richard Payne Knight, to give up his seat as MP for Ludlow and divest himself of responsibilities for running the castellated mansion of Downton Castle which he had had built. He handed it lock, stock and barrel, along with the surrounding ten thousand-acre estate, to Thomas Andrew.

During the previous year Knight had been a moving light in the formation of the Horticultural Society (later elevated to the Royal Horticultural

Society), and when a year or two later he succeeded the Earl of Dartford as president of the Society, he became one of the most influential figures in Britain's gardening world until his death in 1838 – a few days after being elected president for the twenty-ninth time. Men with lesser claims to honour lie in Westminster Abbey. Knight rests with other members of his family in a country graveyard around the church at Wormsley in Herefordshire, a more fittingly tranquil place for a man who was born, and remained throughout his life, a countryman to his bones. He had presided over gardening events in Britain during a period marked by extraordinary improvements of many kinds, and played a full and often leading part in many of them. His position for so many years as president of the Horticultural Society in London, even though it was still a young and immature organization compared to the established academic institutions of Europe, made him enormously influential.

Knight's reputation has faded since his death, but the fruits of his period in office are still apparent in the British countryside in a way he could never have foreseen. In 1824 the plant collector David Douglas was commissioned by the Horticultural Society, while under Knight's presidency, to proceed to the region of the Columbia river, which divides what are now the states of Oregon and Washington. During the next few years Douglas sent back seeds of many of the giant conifers native to that part of the world, which in the latter part of the nineteenth century transformed the appearance of the estates of the landed gentry across Britain. Today these trees pose almost insoluble problems for any film director looking for a location for *Pride and Prejudice*, or any other Jane Austen novel requiring a setting on a country estate. Costumes are in period and styles of carriages, furnishings in houses and every other detail are zealously matched with contemporary fashions. Yet the conifers introduced by David Douglas – nearly ten years after Jane Austen wrote her last novel – steal the show in almost every outdoor shot.

Knight and his patron Joseph Banks were a refreshing change from the cast of characters hitherto concerned with the promotion of botanical studies. With few exceptions the latter had been academics, their livelihoods dependent on the positions they occupied in the universities of Europe. The inquisition of the Catholic Church, epitomized by the trial and humiliating condemnation of Galileo in 1633, had had a distinctly dampening effect on scientific innovation and reinterpretation of the scriptures in southern Europe. The Protestant ethos of the universities in the north was scarcely more tolerant of innovative ideas, and it was still necessary in almost all of them to have been ordained before holding a position of any authority. As such, academics were bound to the doctrines of the Church according to holy scripture. And the description of the Creation in the scriptures explicitly stated that the earth and all things on and in it had been put together in a matter of days, with no allowance for further change. That was a stumbling block highly likely to trip up anyone attempting to hybridize plants (or animals) in order to 'create' new and improved varieties – the orthodox presumption being that if they were not present at the beginning, they could not be summoned into existence thereafter.

Neither Thomas Knight nor Joseph Banks had formal religious connections. Both were men of means who were not beholden to others for their daily bread, and this gave them a freedom of action and capacity for independent thought denied to tenured academics. In addition, Knight spoke to and for others like himself – country gentlemen, farmers and gardeners – and his introductions of improved varieties of strawberries, apples and plums meant much more to them than philosophical deliberations. These varieties were proof that the new science of plant breeding, even if not yet very scientific, produced practical results. His inspiration encouraged others in Britain and all over Europe to follow his example, and to assume the role of creator in their turn.

In 1800, a year after the appearance of Knight's description of his experiments on peas in the Royal Society's 'Philosophical Transactions', a German translation was published in Leipzig. When this translation was read by a man called Christian Andre, it would provide the next link in the chain of discoveries that ultimately revealed precisely how plants transported their genes through seeds. Andre was a leading figure in the Royal and Imperial Moravian and Silesian Society for the Improvement of Agriculture, Natural Science and Knowledge of the Country, informally known as the Agricultural Society.

Andre, a landowner and sheep breeder from Moravia, which is today in the east of the Czech Republic, also quickly recognized the practical applications of Knight's experiments. On his advice, other members of the Society began to use similar methods to raise new varieties of vines and fruit trees. A few months later Andre founded the Brno Pomological Society, where he also started breeding apples with Knight's example in mind.

The next link in the chain was not a university academic, a practical farmer or a gardener. He was a monk called Cyrill Napp, whose wide-ranging interests combined all three. His appointment as abbot of the Augustinian Monastery of St Thomas in Brno in 1827 marked the start of a period of enlightened leadership with extremely productive results. This Augustinian monastery provided exceptionally fertile ground for such activities, but it need not have been so. Other monastic orders were less enlightened, as was pointedly underlined for me during a visit to Brno, a few months before the communist regime in Czechoslovakia was overthrown during the Velvet Revolution.

The Augustinian Monastery of St Thomas had been converted into government offices by the atheist, anti-clerical regime established by the communists, and was not accessible to tourists. Instead, in a bizarrely macabre experience of monastic life, the authorities chose to represent the Augustinian establishment as the Carthusian Monastery, previously inhabited by a less liberally outward-looking, more enclosed community of monks. Visitors to this monastery were greeted immediately inside the front door, by the skeletal, dehydrated corpse of a fully clothed man in an open coffin, who had all too obviously had his throat cut. An elaborate chandelier composed of skulls supported by artfully arranged bits and pieces of human skeletons hung over the coffin – there was no information about the deceased whose skeletons had been elevated in this way. Sconces on the walls held candles within a framework of femurs, tibias and other bones. The remainder of the exhibits dwelt equally heavily on death in one form or another, in preparation for the dramatic conclusion of the experience. A descent into the basement revealed a former abbot, clothed in his habit, lying outstretched on the stone floor, flanked on either side by the desiccated corpses of long deceased monks, waiting in mutual, silent communion for Judgment Day to dawn.

It was not the intention of the authorities in charge of the museum to present monastic life in a favourable light. Nor was this a representative picture of ecclesiastical life in the town as a whole, and especially not of the monks within St Thomas's monastery. Cyrill Napp was a practical man. As abbot, he encouraged the monks to play active parts in the life of the city. They taught in schools and colleges, pursued other scholarly activities and studied the natural sciences, especially the application of science to agriculture and horticulture. Before becoming abbot, Napp had managed a nursery on a monastery farm, where he raised large numbers of seedling apple trees. Shared interests had brought him into contact with Andre, and when Napp was later elected President of the Brno Pomological Society, the two became friends.

Sixteen years after Napp became abbot, a young man knocked on the monastery door, bearing a note introducing him as a potential recruit. The man was Gregor Johann Mendel (11), the son of German-speaking peasants with a small farm in northern Moravia. Unhappy with the life of a peasant farmer, he sought escape by acquiring an education, but proved almost equally incapable of coping with school. Long bouts of depression interrupted his schooling and forced him to return home where much of his time was spent in bed, unable either to contribute to family life or to summon up the effort to continue his studies. He completed the course eventually and moved on to the Philosophical Institute in Olmütz in preparation for entry to Vienna University, but once again his studies were punctuated by long periods confined to bed, unable to face the world. When it became obvious that he was most unlikely to complete the course, a teacher at the Institute who had been a monk in the monastery of St Thomas suggested he seek refuge from his problems in the monastery. He provided Mendel with a letter of recommendation addressed to Cyrill Napp.

Mendel arrived at the monastery as a young man of twenty-one, conspicuously lacking in self-assurance. There he became Brother Gregor and was enrolled by Napp as a student of Holy Orders at the Brno Theological College. Four years later, his problems apparently behind him, Mendel was ordained a priest, and soon took up his duties as a pastor in a city parish, though still based in the monastery. Sadly the respite was short-lived. Despite his obvious intelligence and a personality which clearly appealed to, and impressed, many who encountered him, Mendel proved as temperamentally unfit to be a parish priest as to be a peasant farmer or scholar. Within months he was back in the grip of depression, and had taken refuge in the comforting confines of his bed.

When he recovered Napp sent him off to teach maths, Greek and natural sciences to junior pupils at a school in southern Moravia, and while there to prepare for examination to qualify as a high school teacher. Mendel fluffed his first attempt miserably, but once again Napp stood by him. Despite his poor performance, and the fact that at twenty-nine Mendel was well beyond the normal age for enrolment, Napp managed to persuade the authorities at Vienna University to allow him to enrol as a student. Mendel spent two years there without completing his degree. He studied natural philosophy where he would have been introduced to Koelreuter's and von Gärtner's experiments on plant hybridization, and probably Knight's activities as a plant breeder as well. On his return to the monastery, Napp put him in charge of the garden. He also found work for him as a teacher in the city, standing in as a locum, until he could make another attempt to pass the examination necessary to become a qualified high school teacher. This Mendel failed even more abysmally than the first time and, overcome by nerves and lack of confidence, he again took to his bed.

Cyrill Napp had been a consistent and understanding supporter for some eleven years. Mendel by then was thirty-two, and still so lacking in self-confidence and self-esteem that, despite this support, he had consistently failed every challenge he encountered. He had, however, looked after the monastery garden for a couple of years, and Napp not only allowed him to continue as gardener but also encouraged him by giving him a research project. As a further incentive, he provided Mendel with a glasshouse in support of his experiments.

The project was a study of inheritance in peas. It is not certain the idea was initially Napp's, rather than Mendel's, but that seems highly probable. We cannot be sure why peas were chosen for these experiments, but a copy of the translation of the Royal Society's 'Transactions' in which Knight had described his results lay in the library at Brno and may well have been the inspiration. Napp, through his connections with Andre, would almost certainly have known about Knight's work; we cannot be sure whether Mendel read it, but if he did not, he should have done. Napp could hardly have failed to have considered the possibilities of extending Knight's work during his apple breeding days – and he now recognized in Mendel someone with sufficient education, an interest in maths and an aptitude for gardening well equipped to pursue the study on his behalf.

During his probationary years in the monastery garden Mendel had grown and compared more than thirty different varieties of pea, and had identified seven pairs of contrasting characters that seemed suitable subjects for experimental studies on inheritance. So when he embarked on his experiments in earnest during 1856, supported by the abbot's new glasshouse, he was ready to start making carefully controlled crosses between different varieties straight away. Seven years later, having counted tens of thousands of peas along the way, he successfully completed his experiments on peas, and extended his work briefly to maize, sweet williams, beans and snapdragons.

God now smiled on Gregor Mendel. Not only were peas ideal subjects for such studies, but the characters he selected happened to be well chosen to produce clear-cut, unequivocal results. In addition to that, he appears, almost miraculously, to have escaped serious problems from pests and diseases throughout the course of his experiments despite growing large numbers of pea plants year after year in the same place – normally a recipe for disaster. Heavy infestations of the pea moth, whose larvae burrow into and destroy developing peas while they are still in the pod, would have wrecked his experiments at any time, and he was phenomenally lucky to escape their attentions throughout the course of his observations. As if to underline his good fortune, plants in follow-up experiments in 1864, the year after he had substantially completed his work, were so heavily infested he could obtain no useful results from them.

While Mendel was absorbed in his experiments on peas in the monastery garden and, as far as we know, for the rest of his life, he appears to have been free from the black dog of depression. When the experiments were finished he had not only shown that the characters he had chosen were all passed on from generation to generation, but also that they were inherited according to rules which could be consistently expressed as simple mathematical ratios.

He had also come up with a plausible hypothetical explanation of his results. Mendel postulated the existence of hereditary factors whose presence or absence accounted for the appearance or otherwise of particular characters. The genie was not out of the bottle yet, but Mendel now knew it was there and had started to discern its form. During the remainder of the nineteenth century, scientists in Germany would gradually give form and substance to Mendel's hereditary factors. However, the next century would be dawning before the magic words were discovered which allowed scientists to harness the genie's powers.

Mendel first announced his findings to the world on 8 February 1865, in the course of a two-part lecture to members of the Brno Society for the Study of Natural Sciences. Forty local teachers, college professors and amateur naturalists turned up one cold winter's evening to listen to the first lecture. When he had completed his presentation, replete with mathematical ratios, there were no questions. Four weeks later he delivered the other half with similar results. Most of his audience probably had little idea what he was talking about and failed to see any significance in a rather dull occasion. Later that year transcripts of the lectures under the title 'Versuche über Pflanzenhybriden' ('Experiments in plant hybridization') were printed in the society's official proceedings. Mendel posted copies to botanists in universities whom he thought might be interested. The response could only be described as a deafening silence. Two years later Cyrill Napp died, and in March 1868 Gregor Mendel was elected to succeed him as abbot of St Thomas's monastery. He made no further effort to draw attention to his discoveries and died in 1884. For the remainder of the nineteenth century they remained obscure and largely ignored, though not, as is sometimes suggested, unknown.

Mendel's chief claims to fame lie in what he surmised rather than what he did. They are the product of his imagination and gift for interpreting his observations rather than the originality or conduct of his experiments. The latter built on what others had done before, and their successful conclusion owed more to dogged determination than brilliance of execution, and probably to a little judicious adjustment of the figures. As a mathematician, Mendel must have been fascinated to discover that characters were inherited according to simple mathematical ratios, and immensely encouraged to find these were consistently applicable to a variety of different characters in peas and some other plants – though not quite so consistently as his published data suggest.

But greater and more significant secrets lay behind the façade of the ratios, known to this day as Mendelian ratios in recognition of their discoverer. Mendel's imaginative revelation of such secrets provides the foundation of his fame. He had shown for the first time that characters are inherited from generation to generation according to fixed laws – and not simply inherited, but inherited in logical ways controlled by the presence or absence of actual entities within the cells. Genes were a concept whose time had not yet come, and they remained something beyond Mendel's awareness. However, he did recognize that for characters to be inherited in the way he had observed the embryos must carry factors responsible for their expression, and he conjured up a means of representing these factors that is still used today. He also recognized that there is more to a plant than its appearance may suggest, because qualities can be masked in one generation only to re-emerge in a succeeding one (10). Mendel had opened the way to more informed and logical approaches in breeding new and improved varieties of plants, but nobody paid any attention. Even their discoverer made only the feeblest effort to publicize them, and when this failed he apparently lost interest in them.

While the scientists were peering into the bottle, wondering what the genie could be, contriving ways to learn more about him and devising ways to harness him, farmers, nurserymen, seedsmen, gardeners and others who had formerly regarded seeds as a routine part of the production of cereals and vegetables popped the cork and released him – with wonderful results. They had not waited for Mendel to show them how to become plant breeders, or for scientists to explain the secret ways in which plants passed their inheritance from generation to generation. But, following Knight's example, and no doubt vastly impressed by the monumental output of Karl von Gärtner, they sought to adopt the role of creators and explore previously barely imagined secrets concealed within the enigmatic little bodies of seeds. Until then, new varieties had arisen by chance and their preservation had depended on serendipity. Now they were likely to be the deliberate outcome of someone's attempt to produce a better cabbage, a tastier raspberry or more productive bean. As the nineteenth century ran its course, plant breeders in England, France, Germany and elsewhere in Europe, the USA and even further afield produced an outpouring of new varieties of all kinds of flowers, vegetables, fruit and cereals at an ever-increasing rate. Following empirical rather than scientific methods, they transformed farming and agriculture to hitherto unimaginable levels.

Thousands of amateur and professional plant breeders tried their hand at producing new varieties, covering practically every kind of plant grown widely by farmers and gardeners. The tiny sample introduced below conveys only a hint of the extent of their efforts, and the sea change they brought to both gardening and farming before scientific methods of plant breeding developed during the twentieth century.

Fruit breeders played a consistently prominent part, with eminent firms such as those founded by Thomas Rivers and Thomas Laxton passing from father to son. Rivers raised some fifteen hundred seedlings of peaches and nectarines intended for cultivation under glass, of which some half a dozen were worth naming. His son continued the good work, but was mainly interested in producing new apples and pears, among them the pear Conference, still one of the most widely grown commercial varieties. Laxton produced many new varieties of strawberries, roses, potatoes and peas. Following Knight's lead he also made crosses for experimental purposes between peas – describing the phenomena of dominance and segregation of characters, and also noting patterns of inheritance similar to the ratios found by Mendel. Laxton's sons Edward and William continued well into the twentieth century as notable breeders and growers of new varieties of apples and pears.

In the United States, Charles Hovey, editor of *The Magazine of Horticulture* (for many years the longest running gardener's periodical in the United States), provided active support and encouragement to attempts to breed improved varieties. From the 1840s onwards he was an enthusiastic collector of new kinds of fruit tree, as well as camellias, chrysanthemums and strawberries. Then, practising what he had been preaching, he marketed a strawberry seedling of his own breeding. Known simply as the Hovey

Seedling, it was widely acknowledged as the variety which put the American strawberry growing industry on its feet.

Breeders in the States also paid a great deal of attention to grape vines, attracted by the possibilities of using native North American species to breed varieties adapted to the climate. Hermann Jaeger, born in Switzerland, grew numerous seedlings on his farm on the Ozark Plateau in Missouri in a project intended to find varieties resistant to mildew – in which he was successful. Apart from providing foundation stock for fellow plant breeders, these acquired historic significance when the classic vine-producing regions in Europe were threatened with devastation by the introduction of phylloxera, a bug of American origin to which European varieties had no resistance. Jaeger came to the rescue of the European vintners as they faced the horrific prospect of the total destruction of their vineyards within a few years, by sending them hundreds of thousands of resistant rootstocks on which to graft scions of the susceptible varieties.

Jacob Moore, born in 1836 at Brighton, New York, devoted much of his life to breeding redcurrants, strawberries, pears and grapes using methodical, semi-scientific approaches. These were highly successful, but the poor man suffered severely from unprincipled fellow nurserymen shamelessly pirating his introductions, leaving Moore with very little profit for his efforts. Driven to distraction, he spent much of the latter part of his life trying without success to secure protection for the rights of plant breeders through legislation.

A noted amateur plant breeder from around 1832 onwards was Marshall Wilder, owner of a considerable estate in Massachusetts and one of the leading lights and founders of the American Pomological Society. Pears were his declared interest – as well as camellias and azaleas – but he ranged so widely he can only be described as shamelessly promiscuous. It was said that he went nowhere without a camel-hair paint brush in his pocket, lest he should miss an opportunity to make a cross.

During this time, as a result of the plant breeders' efforts, many of our most popular garden plants became established. In 1850 James Kelway started a nursery at Langport in Somerset, where he specialized in gladioli, then a novelty. (The first large-flowered hybrid had been introduced nine years earlier by Van Houtte.) Using stocks of these, and improving on them with further crosses with various different species, Kelway introduced the first of the modern florists' hybrids in 1861.

The second half of the nineteenth century also saw the development of the hybrids from which modern daffodils and narcissi are directly descended – notably varieties introduced by a Church of England clergyman, the Reverend Engleheart, during the 1880s. Over more or less the same period sweet peas were transformed by Henry Eckford, a Scotsman who, after spending some time breeding verbenas and cinerarias, bought a nursery at Wem in Shropshire. His introductions from here during the last thirty years of the century led to the development of new forms with a wider range of colours and more substantial flowers (12). Sweet peas put the name of Wem on the lips of polite society, and in grateful memory of Eckford the annual sweet pea show is now the most notable and fragrant event on the town's calendar.

Somewhere, someone would try their hand at almost anything, however difficult – even orchids. John Dominy, employed at the famous nurseries of James Veitch near Exeter in Devon, raised the first hybrid orchid from seed in 1856 after crossing two species of calanthe. He passed on his secrets to another of Veitch's employees, John Seden, who had already made a name for himself breeding gloxinias, begonias and other glasshouse crops. During the 1870s and 1880s Seden would raise numerous orchid hybrids from a wide range of crosses, several of which were between species in different genera. Seden's results showed that it was possible to cross even relatively distantly related plant species, established the fascination with orchid breeding among many gardeners and laid the foundations of today's multimillion pound orchid industry.

The prize for the pre-eminent plant breeder of the latter part of the nineteenth century must go to Luther Burbank, who was born in 1849 and lived till 1926. This man, one of the greatest pioneer horticulturalists and plant breeders, produced something like 800 varieties of a vast range of plants, among them apples, pears, plums and strawberries, vegetables of many kinds, garden flowers and cereals – even a spineless cactus for feeding cattle in arid parts of the world, which appears to have become extinct. A few that stand out among his innumerable introductions are the Santa Rosa plum (named after the location of his nursery in California), the Freestone Peach, the foundation of the Californian canning industry, and the Burbank potato. This produced a russet-skinned sport, known as the Russet Burbank. It became the most widely cultivated potato in the USA, and supposedly the only variety acceptable for the production of McDonald's French fries.

The nineteenth century, on an accelerating trend, was the golden age of the nurseryman as plant breeder. They were stimulated by the introduction of plants from overseas, especially those from the 'secret' gardens of China and Japan. Practical farmers and gardeners not only saw seeds for the first time as the means by which hidden qualities possessed by plants could be revealed, but the vigour with which they followed up that realization also exploded notions about the immutability of species forever. From the beginning of the nineteenth century similarities between closely related animals and plants – and the very evident links between many of them, which pointed to evolutionary processes of some kind – were increasingly raising misgivings in the minds of zoologists and botanists about the credibility of the doctrine of divine creation and the immutability of species. But it was one thing to observe and to suspect, quite another to be able to provide scientific evidence for evolution in the absence of any kind of driving force or day-to-day evidence of its occurrence. It was futile to attempt to challenge the entrenched beliefs of those insisting that the biblical description of the Creation answered all such questions, without being able to explain how one species could evolve into another.

The impasse was broken by one of science's more felicitous coincidences. In 1838, just two years after HMS *Beagle* had brought him safely back to England, Charles Darwin read Thomas Malthus's 'Essay on Population', forty years after its publication. Alfred Russel Wallace read the essay sixteen years later. Both were struck by exactly the same thought, recalled by Darwin in a memoir he wrote for his family towards the end of his life. 'It at once struck me that under these circumstances favourable variations would tend to be preserved, and unfavourable ones to be destroyed. The result of this would be the formation of new species. Here then, I had at last got a theory by which to work; but I was so anxious to avoid prejudice, that I determined not for some time to write even the briefest sketch of it.' Darwin's mind had already started to assemble evidence for evolution, despite the greatest misgivings raised by such a challenge to the biblical account of the Creation.

Russel Wallace used almost the same words as those in Darwin's first two sentences to express his reactions, but his response to the revelation was very different. He had become accustomed, through his plant and animal collecting activities and travels in tropical countries, to convert thought into action. Once convinced that only an evolutionary process of some kind could explain what he was continually observing, he sat down, sorted out his thoughts on the matter and set them down on paper. He sent the result to Darwin, who after considerable persuasion consented to make the almost identical conclusions he had reached public in a joint paper presented to the Linnean Society. The publication of Darwin's book, *On the Origin of Species*, in 1859 resolved decades of speculative conjecture about evolution in a way that could not be overlooked. It precipitated the conflict with the ecclesiastical community which Darwin had dreaded, and which he had no stomach to face.

Many members of the scientific community accepted the conclusions reluctantly, and years passed before there was general acceptance of Darwin and Russel Wallace's proposals – especially of our common inheritance with the apes, a highly unpalatable thought to many. The conflict persists to this day, with religious fundamentalists still insisting that no alternative to the biblical account of events is acceptable. In the final years of his life Luther Burbank, incensed by the infamous Scopes or 'monkey trial' of 1925, in which a high school teacher in the USA had to defend himself in court when prosecuted for teaching the 'Heresy of Darwinism', publicly declared himself a free thinker who believed neither in the Creation nor the afterlife. Despite a lifetime spent in practical demonstration of his belief in the mutability of species, this so shocked many of the horticulturalist's God-fearing erstwhile supporters that they inundated him with thousands of letters of protest and outraged comment. Not surprisingly many nurserymen in the States preferred to import new varieties than to tamper with God's creation by indulging in a little creative behaviour themselves.

But even among the clergy, Darwin did not lack influential and effective supporters. On 18 November 1859 Charles Kingsley, rector of Eversley in

Hampshire and author of *The Water Babies*, wrote to thank Darwin for sending him a copy of *On the Origin of Species*. The two men were acquainted through Darwin's sister, who, like Kingsley, was a leading light in the movement to outlaw the use of children as chimney sweeps, a practice that Darwin also found horrifying. In his letter Kingsley wrote with reference to *On the Origin of Species*: 'All that I have seen of it awes me; both with the heap of facts and the prestige of your name, and also with the clear intuition, that if you be right, I must give up much that I have believed and written…I have long since, from watching the crossing of domesticated animals and plants, learnt to disbelieve the dogma of the persistence of species. I have gradually learnt to see that it is just as noble a conception of Deity to believe that He created primal forms capable of self-development into all forms needful *pro tempore* and *pro loco* as to believe that He required a fresh act of intervention to supply the lacunas which He Himself had made. I question whether the former be not the loftier thought.'

Darwin was castigated by religious fundamentalists for challenging the doctrine of the primacy of the Creation and the immutability of species – in other words for his belief that natural selection, based on the differential survival of offspring, was the driving force of change, or evolution. That was ironic since, although natural selection can indeed be the mechanism through which changes occur, it more usually acts as a conservative force, hostile to innovation.

The extraordinarily different shapes, sizes, colours and appearance of dogs to be met on a walk any day in a public park are striking examples of the versatility of the genes within a single species when free to express themselves. By comparison nature seems almost ultimately conservative, so much so that within a single lifespan, especially of a fairly sedentary life, only the most enquiringly observant man or woman would observe any of the clues that point to evolution in action. Natural selection works extremely effectively to suppress variations of the kind that lead to such wide differences under domestication in animals and plants. It is hardly surprising that for so long the biblical doctrine of immutability rather than evolution seemed to those relying on their personal experiences an entirely satisfactory way to explain the world about them.

Natural selection restricts most species, most of the time, to those individuals possessing a narrowly defined range of genes, which in no way reveals the diversity of their genetic types or genotype. Almost all variations on existing genotypes either produce plants less well fitted to survive than established ones, which are immediately eliminated or are at best submerged within a generation or two. The genetic diversity that enables plants to respond to selection is always there, but it is enlisted only in response to some form of change in the *status quo*, or to new challenges that confer an advantage on new combinations. Plants of any given species look much the same – generally speaking at least – and that is why a wildflower is usually easily recognized when we see it, and clearly distinguishable from other species growing with it.

As plants were rooted to the spot, unable to roam in search of a partner, classical philosophers brushed aside the whole concept of plant sex. If plants were unable to move, sexual encounters were clearly beyond them. Two thousand years of intellectual stasis followed, during which flowers were perceived as the epitome of perfectly formed, immaculate beings, untouched by sex. By the end of the nineteenth century – merely two hundred years after Camerarius had dispelled the fiction that most plants were asexual – the processes by which characters were passed down from generation to generation had been elucidated (13). Chromosomes and the location of the genes, Mendel's units of heredity, had been located, though much remained to be discovered about them, while Darwin and Russel Wallace had provided a mechanism for the evolution of plants and animals. Seeds, as the means by which plants transported their genes, had acquired a crucially important role in the breeding and production of plants capable of previously unattainable yields, with undreamed-of qualities and possibilities. The twentieth century would see these promises confirmed and expanded in unimaginable ways, but it was not all to be plain sailing.

CHAPTER THREE

The Making of Seeds

The earliest land plants, such as mosses, liverworts, clubmosses, horsetails and ferns, were spore-producers. Although spores resemble seeds, particularly in their abilities to move and colonize fresh sites, they have distinctively different forms and functions. Spores are not themselves the products of sex; instead, as we shall see, they give rise to stages in a plant's life history that develop male and female organs. Sexual reproduction in spore-producing plants depends upon the presence of water, at least for a short period, to enable male cells to swim to female cells. Despite these limitations, mosses, ferns and their cousins not only flourished millions of years ago, but are with us still – continuing to thrive wherever conditions are to their liking.

A short walk beside the Moeraki river in Fiordland, New Zealand, left me in no doubt how subtle the balance of advantage between the spore-bearing and seed-bearing plants can be. The walk started through a part of forest filled with plants with narrow, grass-like foliage. Astelias, sedges and New Zealand flaxes grew on the ground, perching lilies crowded the forks of the trees, and climbing kie-kies invested tree trunks with a dense cover of narrow, pendant foliage. A couple of hundred metres along the path the scene changed – plants with broad, glossy, deep-green foliage now filled the spaces between the columnar trunks of the trees. There were scheffleras and lancewoods, broadleafs and mahoes, and scarcely an astelia or a New Zealand flax to be seen. A little further on, the black trunks and graceful, spreading fronds of tree-ferns were all around me, and below them the forest floor was crowded with their more earthbound cousins. The fronds of shield ferns, so charged with water that they glistened as though suffused with oil, brushed against my legs, soaking me from the knees down. Colonies of crown ferns, with pale fronds, and the arching fronds of hen and chicken ferns gleamed in the shadow. They lit up dark recesses and small groups of Prince of Wales feathers – supremely stylish Lilliputian tree-ferns, spreading out their fans of lacy, deep green, glistening foliage. Filmy ferns grew among carpets of mosses and liverworts on the lower limbs of trees, reinforcing an overall impression of a more ancient world, constructed before the advent of broadleaved plants.

Yet now these worlds existed side by side, separated by no more than a few hundred metres. The presence (or absence) of narrow-leaved or broad-leaved seed-bearing plants or spore-bearing ferns (14) was due to subtle, almost indistinguishable variations in exposure, soil moisture, past history and present circumstances. Places like this, where one location follows another within metres, emphasize how delicately poised is the balance between advantage and disadvantage. There are many places on earth where few, if any, ferns grow. Elsewhere conditions favour ferns, and there they grow in abundance. Spore-bearing and seed-bearing plants reproduce in fundamentally different ways, but suggestions that one way is better than the other are denied by what we see when we go for a walk in the woods.

Pollen and seeds marked a turning point in the fortunes of plants which eventually gave rise to the vast majority of the trees, perennials, bulbs, annuals and other plants that now share our planet. Nowadays pollen and seeds are found in the gymnosperms, which species includes cycads (15), ginkgos and conifers, and in the flowering plants, the dominant plants of today's world. However, just where and when pollen and seeds arose remains a matter of lively scientific discussion.

The search for answers takes us back some 350 million years to the Devonian Period. Over about 50 million years clubmosses, ferns, horsetails, cycads and, later, conifers defined themselves as separate and distinctively different groups, all competing for space and sunlight. No longer were plants confined, as mosses and liverworts had been, to places where water was readily available. Plants had found ways to survive drought and to thrive in situations remote from wetlands and water courses. They had developed root-like organs capable of garnering mineral resources locked in the soil, and found answers to the problem of reproducing without being able to move around in search of a partner. In doing so they had developed structural and water-conducting systems, leaves and roots, specialized sexual organs, spores, seeds and pollen – all in forms familiar to us today. By developing woody tissues, plants could raise themselves high into the air in the competition to overshadow their fellows. At one time or another, and in one place or another during this and the succeeding Carboniferous Period, forests of tree-ferns, clubmosses, horsetails, gymnosperms and plants of many kinds long gone filled the valleys and covered the slopes of hills and mountains. By around 300 MYA the only major innovation left for plant life on land was the evolution of flowers. It would be another 150 million years before flowers appeared.

The Devonian ferment would appear to be a promising source of answers to questions about when, where and how seeds originated. Unfortunately plant fossils of diverse and varied forms appear together in similar locations, with very similar dates. Occasionally fossils are found that reveal the existence of plant forms unknown today, some of which have been hailed as missing links between spore-bearing and seed-bearing plants. They include a group of fossil plants known as the progymnosperms, once proposed as direct ancestors of cycads and conifers. They flourished early in the Devonian Period and, geologically speaking, became extinct not long after (about 50 million years later). Ideas that progymnosperms might be the ancestors of seed-bearing plants were challenged when fossils of a plant called *Runcaria* were reinterpreted in 2004. *Runcaria* was first described from 1968 from 385 million-year-old fossils discovered in Belgium and is contemporary with the progymnosperms. *Runcaria* has features specific to seed plants and has been interpreted as a seed-precursor: a member of the seed lineage leading to seed plants. At the dawn of the twentieth century there were other candidates that appeared to link spore-bearing and seed-bearing plants. The case for seed ferns, plants with a superficial resemblance to cycads, being the link was abandoned when more was discovered about how seed ferns reproduced, and when it became clear that the first seed ferns appeared long after seed-bearing plants. Further evidence is needed to answer these questions on the origins of seeds.

All land plants have two phases (or generations) to their lives, and they alternate between them during their lives. One phase produces sex cells (so-called gametes) and the other phase produces spores. These phases are known as the gametophyte (literally 'gamete plant') and the sporophyte (literally 'spore plant') respectively. The process is called the 'alternation of generations'. Imagine the fern sitting in a pot on your windowsill or growing in a damp wall: it is a sporophyte. Spores are created by spore-producing structures in brown areas on the backs of the leaves. They are blown away in the wind, and when they land in a suitable area spores germinate to produce gametophytes. Fern gametophytes are short-lived, fragile plates of green tissue, called prothalli (prothallus in the singular form). The sole purpose of the prothallus is to produce and support the male (antheridia) and female (archegonia) organs. Motile male cells (sperm), produced by the antheridia, migrate to the archegonia in a film of water, where they fertilize a single egg cell. This fertilized egg cell then develops into a new sporophyte. As the sporophyte grows, the prothallus (gametophyte) shrivels and dies, and the process can start once again.

In seed plants, the gymnosperms and flowering plants, the gametophyte stage of the life cycle is minute and supported by the sporophyte; it does not have a separate existence, as in the case of the fern. In the majority of land plants, the sporophyte is the dominant phase. However, in mosses and liverworts, the green, leafy stage, with which we are familiar, is a gametophyte; the sporophyte is the capsule that sticks up from the green leaves in autumn. The difference in size between these two phases gives us potential clues as to seed origins.

Imagine a scenario where a fertilized egg cell did not develop immediately into a new sporophyte, supposing instead that protective walls grew around the embryo. To all intents and purposes this would be a seed – and indeed the technical definition of a seed is an integumented, female sporophyte. The prothallus would shrivel and die and the 'seed' would be left behind. Within the 'seed', the embryo would be protected until favourable conditions provided the opportunity for it to grow.

The next step would be to do without spores at all, and instead produce fertile organs similar to those produced by a prothallus on modified fertile shoots. The egg cells, produced in archegonia, could be fertilized by male cells, from antheridia, carried by the rain or, in the course of time, by wind, fulfilling the functions of pollen. If the fertile fronds were multiplied, stacked up one above the other on short stems and became horny or woody, the result would be cones – like those found in cycads and conifers. The progression from a spore-bearing to a seed-bearing plant is neither improbable nor particularly complex. It just happens to be obscure because so few fossils tracing its course have yet been found.

Seeds inherit qualities from both their male and female parents: each seed is thus genetically distinct from both its parent and its siblings. Seed-bearing plants are conspicuous and pre-eminent in our daily lives and in our gardens. When we encounter mosses, ferns and other spore-bearers they are usually in specialized habitats, often nooks and crannies, and places that for one reason or another are less attractive to the seed-bearing gymnosperms and flowering plants.

The apparent evolutionary advantages of seed-bearing plants may be no more than a reflection of climatic conditions during the past few hundred million years. We happen to live in an era in which flowering plants dominate the world's terrestrial vegetation. That has not always been so, and it may not always be so in the future. One day our planet may return to the wetter, milder conditions that favour ferns and other spore-bearing plants. Spore-bearing plants do things differently. This does not make them less effective; after all, the world's coal deposits are the remains of giant, tree-like club mosses and horsetails.

The earliest cycad fossils date to some 280 million years ago and are among the earliest seed-bearing plants still with us. Once abundant, they are now seldom more than a scattered, invariably characterful presence. Yet there are places where a taste of their ancient ascendancy can still be experienced. One is in the Modjadji Nature Reserve, the Domain of the Rain Queen, in South Africa's Northern Province. Here cycads congregate so densely over so large an area that they form a veritable forest.

It is an extraordinary place where dark, harsh, spiky, primeval plants stand gaunt and crowded in the heat. Broken fronds, hanging loose and scraping together, sigh and clatter, rustling insistently in the warm breeze, as they have sighed, clattered and rustled for millennia past. Like aloes or palms, these plants are altogether more elemental leftovers from a time unimaginably distant. Their scaly, fibrous trunks are blackened by long dead fires. Their fronds, cut jaggedly from some thick, insensitive substance daubed a dark, uniform, burnished bronze and green, bear little resemblance to the amenable materials from which leaves are made. Each is a composition of discords. Some are upright, others writhing like the bodies of giant reptiles. Yet others have bulbous, lumpen, improbable-looking excrescences bursting from twisted limbs. Grotesque, yellow cones erupt gracelessly, awkwardly posed and angular amongst the fronds.

These hulking, misshapen, angular, metallic monsters were among the early inventors of pollen and seeds. Inventions which, hundreds of millions of years later, would be inherited by flowering plants, and lead to a mind-boggling array of alliances between plants and animals. Pollen transformed sexual reproduction; wafted from plant to plant by the wind, it enabled matchmaking between distant partners. Seeds, blown by the wind, made it possible for offspring to seek out and colonize new places away from the overbearing presence of their parents.

The cycads in the Modjadji forest are of two kinds. Not two different species, but two different sexes: they are dioecious. Male trees produce pollen, and females carry the cones within which the seeds develop. This seems straightforward enough, but such simple division of sexes is unusual in the plant world. The alternatives available to plants, and the arrangements of their sexual organs, enable them to express their sexuality more imaginatively and in more varied ways than animals. In the great majority of animals boy meets girl, and girl becomes pregnant. No boy, no pregnancy – apart from the phenomenon of virgin birth, most familiar to gardeners among greenfly. (In such circumstances every offspring is a clone of its parent, and inevitably female.) Cycads are by no means alone in their segregation of the sexes. Gardeners may learn to their cost that they must purchase two holly trees if they want berries – one female, the other male.

They may be surprised, however, to learn that 'Golden King' provides berries and 'Silver Queen' the pollen.

Ginkgos, the last representative of an otherwise extinct group of plants, are dioecious. So are several major groups of gymnosperms, notably members of the yew family and the huge family of the podocarps hailing from the southern hemisphere. Dioecy is widespread, but not common, among flowering plants, and trees and shrubs from the southern hemisphere appear more inclined to separate the sexes than those in the north. The obvious biological disadvantage of dioecy is that only females bear seed so, as the gardener discovers with hollies, half the population is biologically unproductive. Each female has to produce two offspring capable of growing to maturity and reproducing instead of the single success sufficient for plants that are both male and female (hermaphrodite).

Most plants are hermaphrodite, but it does not follow that most plants are self-sufficient seed producers. Nature abhors self-pollination, going to considerable lengths to ensure that cross-fertilization is the rule. Much ingenuity has gone into devising ways of ensuring that pollen grains transferred from flowers growing on a neighbouring plant are more likely to produce seeds than pollen from the same or another flower on the same plant (16).

Self-pollination is most frequently discouraged by arrangements whereby anthers produce pollen when the stigmas are not receptive. Usually anthers are mature before the style, but sometimes the sequence is the other way round; the styles have shrivelled and the stigmas become non-receptive before the plant's own pollen is released. Often anthers and stigmas occupy positions which reduce the likelihood of direct pollen transfer. They may grow at different levels, so there is no possibility of direct contact, or anthers may face outwards away from the stigmas.

In orchids there is a double safeguard. The anthers and stigmas mature at different times, but because the anthers lie behind the receptive part of the stigma pollination is possible only when an insect, probing into the recesses of the flower, bodily removes the anthers and transports them to another flower. Next time you find an orchid push a matchstick into the flower, imitating the probing movements of an insect. When you remove the match you will find, if you have been lucky, an entire stamen, like a little pollen-packed balloon, stuck onto the matchstick. Walk around for a minute or two, then push it into another orchid flower; you are now playing the part of an insect. You will notice the balloons have bent over so the anthers point forwards so that they are pressed against the shield-shaped stigmatic surface as they enter the flower. The anthers will stick, releasing tens, or even hundreds, of thousands of pollen grains to fertilize the thousands of egg cells lying in the ovules of the orchid flower (17).

Konrad Sprengel was not the first person to look into primrose flowers and wonder why there were two subtly different models – but he was the first to write down what he saw. In some flowers, the styles extend to the mouth of the tube formed by the bases of the petals, while in others they scarcely reach halfway up. Sprengel could not explain this, and it was left to Charles Darwin to do so.

Darwin was one of those people for whom the natural world was a source of constant fascination. His most famous work, *On the Origin of Species* (1859), set out ideas of evolution that involved variation within species, the inheritance of characters, selection, time and adaption, and developed them to explain the diversity of life. Yet he also published other important books, including *The Effects of Cross- and Self-Fertilisation in the Vegetable Kingdom* (1876). In this book Darwin described flowers with long styles as 'pin-eyed' and those with short styles as 'thrum-eyed'. When he cross-pollinated one with another, he obtained seeds only when he pollinated thrum with pin, or vice versa – pollen from pin to pin or thrum to thrum produced no seeds. This simple device of Nature's ensures that primroses are almost invariably cross-pollinated. A similar (but more complex) arrangement is found in the flowers of the purple loosestrife. In these flowers styles may be long, medium or short, and the anthers may be on long, medium or short filaments.

Cycads harnessed the winds. Conifers did too, and many of our most familiar broad-leaved trees and shrubs still shed their pollen into the air, banking on the chance that enough of it will come to rest on the stigmas of neighbouring flowers to produce sufficient seeds to maintain the population. Not surprisingly, such plants have to release enormous quantities of pollen and produce their flowers in readily accessible locations to increase the chances of a hit. Birches, hazels, ashes and many other woody plants produce flowers on bare branches before their leaves develop. Grasses carry their flowers in dense flower heads above the level of the leaves. Winds are oblivious to fragrance or the sweet attractions of nectar, and colourful petals merely obstruct access of pollen to the stigmas. Wind-pollinated flowers are often stripped to the bare essentials. For example, catkins are composed of tiers of densely packed flowers that are little more than anthers or pistils. Pollen, floating freely in the air, comes to rest haphazardly, and an individual stigma is unlikely to be the resting place of more than a few pollen grains. So, as a rule, each stigma serves a single ovule. A grass seed is the fruit of a single flower; a birch flower has two styles, each of which services one of the pair of ovules within the ovary. Female flowers on an oak tree are carried in groups of one to three, and produce one to three acorns. The scales which compose the cones of conifers form a series of pollen traps, each of which produces just two seeds.

Wind serves plants well as a distributor of pollen, but its haphazardness leaves obvious possibilities for improvement; not even cycads have remained wholly loyal to the wind. Small beetles foraging among the male cones of South African cycads pick up pollen and fly off to female cones, carrying the pollen with them. Fertile seeds are produced as the incidental result of foraging, rather than a contrived partnership between plant and beetle.

Unlike the wind, insects do not work for nothing, nor do they distribute pollen haphazardly. Insects serve only those plants that attract their attention, provide them with somewhere to land when they arrive, guide them in the direction they should go and reward them for their efforts. Attraction, a resting place, guidance and reward make up the brief for the construction of a flower.

Fragrance and colour are the sirens and petals the landing platforms. Shadings, lines and scribbles on the surface of petals, often invisible to our eyes until revealed in ultraviolet light, are guides, while the rewards, of course, are pollen or nectar. Put it all together and the making of seeds becomes the *raison d'être* for the making of flowers. Everything that gives us pleasure – fragrance, colour, attractive markings and nectar from which the bees make honey – is there solely to make it more likely that pollen from one flower will be carried to the stigma of another.

Flowers reward bees with nectar or pollen, and often both. Clover, apple, borage, heather and lime provide nectar and all are famous sources of honey. Goat willows, poppies, old man's beard and St John's worts provide a pollen reward. Whether a bee visits a flower in search of pollen or nectar, flowers ensure that it is bound to come away dusted with pollen – the toll that flowering plants pay for better prospects of pollination and the production of more seeds. Most produce pollen profusely to ensure that a tiny proportion is carried by an insect or other pollinator to a receptive stigma.

Animals carry pollen from flower to flower in far greater quantities than the bare minimum needed to fertilize the ovules contained in the flower's carpels. Unlike wind-pollinated plants – in which an individual stigma is unlikely to receive more than a few pollen grains, followed by the production of a single seed – the stigmas of tomatoes, poppies, foxgloves, campanulas, orchids and many others often receive many hundreds of pollen grains. And, in due course, their flowers produce correspondingly large numbers of seeds. Nevertheless, numerous other flowers, including members of the daisy family, strawberries and potentillas, which are also pollinated by insects, stick to the one stigma/one ovule/one seed rule observed by those dependent on the wind.

Nectar, unlike pollen, plays no part in the life cycle of the plant, costs little to produce and evolved early in the evolution of the flowering plants as an economical way to attract and reward animals. It is secreted in nectaries, or specialized glands. Nectaries appear in so many different forms and parts of the flower that early botanists were at a loss to determine their function. The buttercup family is particularly diverse in this respect, and in one species or another every part of the flower plays its part as a nectary. They appear as modified sepals in monkshoods, globe flowers and hellebores. Pasque flowers sacrifice a few stamens by converting them into nectaries, and marsh marigolds locate them in small depressions on the sides of the carpels. Wild clematis, or old man's beard, dispenses with them entirely and relies on pollen to attract bees and flies, while meadow rue – lacking nectaries – dispenses with insects, relying instead on the wind to carry pollen from flower to flower. The nectaries of buttercups are at the base of their petals, while in columbines they are tucked away in petal spurs, where they are accessible only to insects with long tongues. Bumblebees, whose tongues are far too short to reach the nectar, are nectar robbers, however, and collect the nectar by biting through the base of the spur.

But flowers also cheat. The beautiful flowers of the grass of Parnassus, which resemble elegant, ivory-coloured buttercups (though they are no relation), are both deceivers and rewarders. Instead of producing pollen, five out of the ten stamens are tipped with golden glands which look just like gleaming drops of nectar, but offer no reward. The nectar reward (an energy-intensive substance for the plant to produce) is to be found at the base of these sterile anthers. Ploy and counterploy are the outcome of millions of years of evolution – but, as in human societies, rogues will always bend the rules to make profits.

Confusion about the functions of nectar reigned among academics until early in the nineteenth century, and speculation gave rise to a remarkable variety of theories. Some thought it caught and absorbed pollen, and so played some part in the fertilization of the ovules. Linnaeus observed the nectaries, but did not propose any function for them. Some regarded nectar as an excretory product; others concluded that it was actually harmful to plants and that the bees provided a service by taking it away. Some thought it was a way of using up excess energy, later to be devoted to nourishing the developing ovules. Sprengel was the first to state unequivocally that its function was to attract bees in order to pollinate the flowers, but, as we have seen, nobody paid him much attention (p. 47).

It seems scarcely credible that the link between insects, especially bees, and nectar could have remained so obscure until well into the nineteenth century. The role of pollen in plant reproduction appears to have been better understood by gardeners, few of whom committed their ideas to paper, than by academics, who were all too eager to promulgate their opinions. Perhaps bee-keepers, guided by a practical rather than a philosophical approach to the world around them, were better informed about the nectar-gathering and pollinating activities of honey bees than their contemporaries in the universities.

Nectar, pollen and the honey bee play such fundamentally important parts in the making of seeds, and in the wellbeing and prosperity of human beings, that it is easy to overlook the roles played by all the other animals enlisted by plants to transport pollen from flower to flower. Insects are pre-eminent, and the innumerable special relationships that exist between particular insects and particular flowers have filled volumes. As with nectaries, even closely related flowers may be specialized to attract different pollinators or groups of pollinators, and the shape, colouring and fragrance of the flowers they produce tell us a great deal about the animals that they enlist as pollinators.

The colourful, broadly trumpet-shaped flowers of rhododendrons make them some of the best known of all garden shrubs. The abundance of seeds they produce is a tribute to the effectiveness of the bumblebees and honey bees that pollinate most of the widely grown garden rhododendrons. However, a group known as the vireyas depend on pollinators other than bees. Unlike the familiar terrestrial rhododendrons, many vireyas grow epiphytically on the branches of trees in relatively cool, moist conditions in the tropics. They live among the forests on the mountains, extending south across an enormous expanse of eastern Asia from northern India, through Burma and down the Peninsula Malaysia; they are especially prominent throughout the Indonesian Archipelago, and exist in astonishing numbers in New Guinea. One species has even crossed the Timor Sea to colonize mountains in northern Queensland, Australia.

The flowers of vireyas vary more than those of the familiar, hardy garden rhododendrons; and their colours and forms provide clues to the pollinators on which they rely to fertilize them. Narrowly tubular flowers, often hanging down and coloured red, purple and sometimes pink, are pollinated by birds. Attracted by these colours, birds plunge their beaks deep into the flowers to sip the nectar secreted at the base of the tubes, picking up a dusting of pollen on their foreheads as they sup. Broadly expanded, shallow, cone-shaped flowers, yellow or pink in colour, provide platforms on which butterflies settle. Others, with narrowly tubular flowers of white, cream or pale yellow, release their fragrance as night falls to attract moths, which insert their long proboscises into the hearts of the flowers as they hover in front of them. The shallow, bowl-shaped, fragrant white flowers of other vireya flowers provide feeding stations for bats.

Insects of many kinds, birds, mammals, even reptiles have been inveigled, bribed or shanghaied to pollinate plants in one place or another. Monkey flies with exaggeratedly long proboscises pollinate the flowers of bulbous plants in South Africa. Moths in Britain seek out the flowers of campions, laying their eggs where the larvae can hatch out to feed on the developing seeds and pollinating the flowers while doing so. The plant sacrifices a proportion of its seeds, but evidently the benefits of a dedicated pollinator more than make up for that. Minute wasps, imprisoned within immature flower heads, pollinate figs as the price of their release, and wild arums and Dutchman's pipes entrap small flies and beetles – releasing them later with a dusting of pollen to transport to another flower, where they are imprisoned yet again before being eventually released.

Some plants are specialists, forming exclusive alliances with a single species of insect. Others, such as the carrot family, provide an open table for all comers. Their broadly spreading, flat heads, composed of hundreds of small, white flowers, are accessible to a huge variety of flies, bees, wasps, butterflies and beetles. Sunbirds in South Africa and honeyeaters in Western Australia pollinate the flowers of banksias, while pygmy possums are attracted to the nectar produced in the flowers of low-growing banksias and dryandras. Sunbirds pollinate proteas in South Africa, and rodents in search of the rewards of nectar perform the same service for African lilies. In East Africa, giraffes, browsing the tops of thorn trees, are believed inadvertently to distribute pollen from flower to flower (18).

Over the millennia networks of alliances have developed, and in places almost every species within a genus of plants may have its own pollinator. Western Australian smoke bushes, of which there about fifty species, produce clouds of whitish flowers – not unlike smoke when viewed at dusk with the light behind them. Each species has evolved a special relationship with a particular species of primitive bee. The bees appear to be attracted by the particular shape of small, black bodies among the flowers, perhaps because they mistake them for fellow bees. One species of bee pollinates one kind of smoke bush; each species of smoke bush exists in strict genetic isolation from neighbouring members of the genus. Or they used to do so, until the arrival of honey bees, to whom a smoke bush of any kind is an equally acceptable source of pollen or nectar. As the honey bees move indiscriminately from flower to flower, promiscuous pollination leads to the appearance of hybrid offspring – to such an extent there are fears that in places some species will become extinct.

The sole purpose of all this ingenious effort is to provide a carrier service able to deliver pollen to the receptive stigma of a flower. Pollen grains are the vegetable equivalent of plastic carrier bags; they serve the most ephemeral of purposes, yet are composed of one of the most persistently non-biodegradable substances known – a substance called sporopollenin. This resists decay so effectively that pollen grains deposited thousands of years ago from hazels, oaks, alders, birches and other trees are the principal means of finding out which trees grew where in the distant past. Pollen grains look to the naked eye like minute, yellow spheres, with no particularly striking characters. Under the microscope, however, these undistinguished yellow spheres are shown to have a great variety of shapes with distinctively different, sculptured and incised patterns on their surfaces, such that genera and even species can be distinguished.

After pollination, pollen grains are left piled up on the stigmatic surface – standing at the front door, so to speak. They still have to gain entry and, having done this, to transfer the chromosomes contained in their nuclei to egg cells in the ovules in the depths of the flower. The first barrier to entry is possession of the right password. The door opens only for pollen grains of the appropriate species – not exclusively, otherwise hybridization would be impossible, but at least only for those related sufficiently closely to provide acceptable credentials. The next stage varies, depending on the plant concerned. The front door may be only millimetres away from the egg cells. The stigmatic surface in poppies, for example, is a broad platform immediately above the placenta, around which the ovules are clustered. In contrast, pollen grains deposited on the stigma of a hibiscus flower still have some six or seven centimetres to go before they reach an ovule.

The spermists (p. 35) assumed, in the absence of any empirical evidence, that the entire pollen grain plunged down the style which, despite appearances, they assumed to be hollow, until it came to rest within the ovule. Joseph Koelreuter's work in the 1760s put paid to that idea when he observed that instead of plunging bodily through the stigmas, pollen grains lay around on the stigma surface and then, apparently, disintegrated. Koelreuter's microscope failed to reveal root-like growths (pollen tubes) emerging from the grains and growing down the centre of the styles towards the ovules. He was also unable to observe that the tissues within the styles were composed of open, spongy cells that allowed the pollen tubes free passage on their way to the ovules, and even provided them with nutrients.

Journey's end for a pollen tube is the entrance to one of the embryo sacs in the egg cells within the ovules. It is not an easy task – in a race where winners take all, hundreds of pollen tubes may compete to get there first. Their job is done when one of the two nuclei they carry enters the embryo sac and fertilizes the egg cell. The fertilized egg cell divides and a new embryo is born, the inheritor of qualities contributed by the male and female parents.

Joseph Koelreuter, lacking a sufficiently powerful microscope, was unable to observe these events. Konrad Sprengel, if told about them, might have accepted the account as the logical extension of his findings. Gregor Mendel would have felt the story was less than complete. He might have nodded assent at the first time of hearing, but may, on reflection, have sensed something missing. For Mendel had proposed that every plant inherits two sets of hereditary principles: one from the pollen parent (male), the other from the stylar (female) parent. At first sight that proposal appeared to square with the idea that a nucleus from the pollen grain fertilized a nucleus in the egg cell; evidently that was the way the hereditary principles were transmitted to the offspring of the union. However, in the next generation the pollen nucleus, already the inheritor of two sets of hereditary principles, would combine with a similarly endowed egg cell, which makes four sets of hereditary principles. And it gets worse. The offspring of the following generation would be endowed with eight sets, the next sixteen, and then thirty-two and so on *ad infinitum*. Simple arithmetic reduced the proposition to an absurdity. Somehow, generation by generation, common sense insisted there had to be some way by which two sets of hereditary principles were reduced to one before the pollen nucleus next met the egg cell nucleus.

One of those who contributed to finding the answer to the conundrum was Eduard Strasburger, professor of botany at Bonn University. Strasburger was the most prolific and influential German botanist during the latter part of the nineteenth century. The first edition of the textbook he compiled, *Lehrbuch der Botanik für Hochschulen* ('Textbook of botany for universities'), appeared in 1894. It has become a botanical text of such eminence that the thirty-fifth edition was published in 2002.

Strasburger was born in Warsaw in 1844 and, after attending the Sorbonne in Paris, he moved to the University of Bonn. He then worked first as an assistant and then as a professor at the University of Jena, before returning to Bonn in 1881. The professorial establishment of the German universities included many of Charles Darwin's most fervent supporters. Strasburger's interest in how cells worked, and particularly the role of the nucleus, involved him directly in questions of sex and heredity in plants, and he was among those who accepted and promoted Darwin's ideas.

At Jena, Strasburger experimented with ways to use the newly discovered synthetic dyes to stain the walls and contents of plant cells (19). Until these became available, it was extremely difficult for plant scientists studying the cellular structure of tissues under the microscope to discern or identify what they saw. Sections of plant stems or developing flowers appeared to be composed of amorphous masses of tissues with shadowy, indistinct features that could scarcely be made out, much less identified. When sections were infused with one or another of the dyes, however, different tissues took up dyes to different extents, and what had been amorphous took form and shape. Cell walls became defined as clearly as if they had been outlined in ink. Bodies within the cells, previously grey shadows swimming in a watery grey porridge, acquired substance and distinctive characteristics. The new science of cytology, the study of the structure and function of cells, had been born. After Strasburger's return to Bonn, the Botanical Institute under his leadership became the world leader in cytological studies.

The nucleus had long been known as a consistent presence in almost all cells, and it was one that attracted great interest. Yet for all the speculation about its function, its ill-defined, shadowy presence in unstained preparations provided little scope for informed opinions. The use of stains changed that for ever. Cell nuclei became well defined objects with features which could be readily observed, and they immediately attracted the interest of Strasburger and his students. Convinced by his observations that the nuclei played central roles during the formation of new cells, he concluded that they were the carriers of essential information that enabled cells both to function and to contribute in an orderly way to the formation of new tissues.

Meanwhile a contemporary of Strasburger's, Walther Flemming, holder of a chair at the Anatomical Institute of the University of Kiel, was studying the behaviour of nuclei and observed thread-like objects within the nuclei with a particular affinity for certain dyes. In 1888 another German, Heinrich von Waldeyer-Hartz, gave these the name we know them by today – chromosomes, literally 'coloured bodies'. The carriers of Mendel's units of heredity had been discovered, and named. The discovery of the units themselves, which we know as genes, lay in the future.

With the aid of the new stains, Strasburger and Flemming were able to observe precisely how chromosomes behaved during cell division. They made sections of tissues in which there were numerous actively dividing cells, the tip of a root for example, and observed chromosomes at every stage of the production of a new cell. What they saw was a revelation. The story began with a gathering of the chromosomes around the centre of the cell and continued as each divided into two to produce two identical sets. It then entered its final stages as the sets separated, by moving to opposite ends of the cell. The operation was completed by the formation of a wall between the two sets of chromosomes – thus making two cells where one had been before. Simple division of this kind, known as mitosis, is the means by which the genetic constitution of a plant is passed on from cell to cell. It ensures that, apart from rarely occurring mishaps, the nucleus of every cell – whether in the roots, shoots, flowers or leaves – contains precisely the same information.

Every cell, that is, except those in pollen grains and the egg cells within the ovules, and this holds the answer to the conundrum that would have left Mendel so puzzled. The cells from which pollen grains and egg cells are produced divide in a fundamentally different way that involves two phases. In the first phase, pairs of chromosomes, derived originally from the male and female parents, move to lie alongside each other in a plane across the middle of the cell. Here they exchange sections of chromosome, in what appears to be a random fashion, before separating into two groups – each with a complete set of chromosomes drawn haphazardly from the male and female representatives. One set moves to one end of the cell, the other to the other, and a new wall forms between them to make two cells. In the second phase, the chromosomes of each cell gather around the centre of the cell, divide into two to produce two identical sets, and finally move to opposite ends of the cell. The products are four cells, each with one set of chromosomes. This process, known as meiosis or reduction division, not only reduces two sets of chromosomes to a single set before pollen cell meets egg cell, but also, by mixing and scrambling the chromosomes, ensures that each and every pollen or egg cell ends up with a unique mix of genes, derived from maternal and paternal chromosomes.

But is all this dancing with chromosomes really necessary? How much simpler it would be if the mother cells, which produce the egg cells, dispensed with the rigmaroles and uncertainties of waiting for pollination and fertilization and grew into embryos without undergoing reduction division. Yet all but a few plants favour the complications introduced by sex. Why should sex be so important in the making of a seed? The answer is that it creates offspring composed of mixtures of their parents' genes in innumerable different combinations, and this diversity is the key to survival when disasters strike or new opportunities arise. Sexual reproduction, and the genetic turmoil which accompanies it, is the insurance policy that safeguards the long-term survival of a species by ensuring a mixture of genetic types (genotypes), some of which may be able to seize the moment.

Like all insurance policies, premiums have to be paid. In the short term such premiums represent unrewarded costs and, just like humans, some species have opted not to pay, or to pay at reduced rates. They work on the principle that a plant which has grown up, flowered and produced seeds is a living demonstration of the possession of an appropriate genotype for the location. Exposing a plant's genes to the melting pot of random mixing with those of another plant produces numerous variations, some of which will be ill-matched to the needs of the location. A plant can hardly do better for its offspring than pass on its own genotype. However, even in the short term, this is a dangerous strategy and one that leads almost inevitably to disaster. Plants are constantly exposed to infestations of pests and fungal infections, and genetic diversity is their major weapon in repelling these invaders.

Some species do produce embryos and seeds without fertilization, and almost all the siblings produced will be clones of the mother plant. Most dandelions do this, as do species of whitebeam, blackberry and hawkweed. The phenomenon is known as apomixis, and gardeners have virtually ignored the opportunities it offers. If we could find ways to make it happen to order, apomixis could be a most useful way to propagate many plants. However, at present, to get genetically identical copies gardeners must propagate them vegetatively from cuttings, divisions or grafts. It is a waste of time to sow seeds of almost any particularly desirable plant with the expectation that it will grow up in the image of the chosen plant that produced them; sex ensures that this will not happen. However, if we could persuade plants to produce seeds apomictically, every seedling would be a member of a clone, and each seed a replica of the plant that produced it.

Cultivated plants escape many of the demands to pay insurance premiums, and after becoming domesticated many of them greatly reduce the payments. These include lettuces, French beans and tomatoes, peas, wheat, rice and barley, all of which are consistently self-pollinating – to the great benefit of those who cultivate them. Farmers and gardeners approve of self-pollinating plants because such plants are naturally inclined to uniformity and lend themselves to the production of new varieties. Because they are self-pollinating, a lettuce with particularly crispy leaves or wheat that produces ears with more and larger grains than its fellows is likely to produce seedlings with the same qualities. New varieties, in which the desirable qualities are present in every seedling, can be produced within a few generations by collecting seeds, sowing them and removing any plants that do not possess the qualities desired.

Maize, onions, carrots, runner beans, cabbages and numerous other cultivated plants never surrendered their aversion to self-pollination; they follow nature's preference for cross-pollination. Until the 1950s individual plants of a variety of Brussels sprout or maize were quite variable. Some plants grew larger and more vigorous than others; some cropped a little earlier and others later. There would be variations in the colour, texture or other qualities of their seeds or fruits. This independence and tendency to vary was curbed by the invention and introduction of F_1 hybrid seeds.

The story behind F_1 seeds started two hundred years ago when Thomas Andrew Knight chose peas for his experiments on inheritance. When he crossed a tall variety with a short one, he noticed that all the offspring, of both the tall and the short parents, were notably similar one to another; they grew even taller and more vigorously than their tall parent. Peas have responded to cultivation by becoming self-pollinating, and so Knight's crosses were made between two habitually inbreeding, homogenous varieties. The extra size of the offspring was an effect now known as hybrid vigour, and their uniformity was also a characteristic feature of the first generation (the F_1) of a cross between two inbred lines.

The production of F_1 hybrids from cross-pollinating parents, such as maize, is rather more complicated, but also more rewarding, than making hybrids between peas. First, inbred lines are produced artificially by self-pollinating plants for several generations against their natural inclinations, during which they decline in vigour generation by generation. Then pollen from one line is used to fertilize ovules in the cobs of the other to produce hybrids between the two. This releases their pent-up hybrid vigour, and also produces a generation in which every plant is closely similar to its siblings. Unfortunately, the effect is for one generation only, and F_1 seed can be produced only by constantly recreating the inbred parent lines – the cause of its high cost.

Mendel surmised that plants contain two sets of hereditary principles – one from each parent. Simple cell division (mitosis) passes on two complete sets of chromosomes to each new cell; it is the default condition, producing a cell described as diploid. Some plants have more than two sets of chromosomes and are known as polyploids. Depending on just how many sets they possess, the plants are labelled tetraploids (four sets), hexaploids (six sets) and octoploids (eight sets), while some ferns have over eighty sets of chromosomes. Polyploids with odd sets of chromosomes, for example triploids (three sets), tend to be sterile. Polyploids tend to be larger and more luxuriant in their growth, and to produce juicier fruits than diploids. As a result, a significant proportion of our cultivated plants are polyploids.

14 The crozier of a developing fern leaf is a characteristic feature of these spore-bearing plants. Ferns are typically plants of damp areas and are rather small, although some take on tree-like forms.

15 Cycads are seed plants of the tropics and subtropics that have a large crown of leaves and a stout trunk. Furthermore, cycads have male and female cones on different plants.

16 The flower of the poppy shows the ring of black, male anthers, liberating their pale coloured pollen, surrounding the female carpels. The spoke-like ridges of the stigmatic surfaces are where pollen eventually germinates so that the pollen tube can fertilize the ovules inside the poppy ovary.

17 The flowers of orchids are highly elaborate and have attracted the attention of generations of gardeners. However, evolutionarily they show considerable evidence of co-evolution with their insect pollinators to ensure that pollen is transferred from one individual of a species to another.

18 *Opposite* Thorn trees are emblematic of the African savannas. There is evidence that some acacia species are pollinated by a most unlikely animal – giraffes. Acacia flowers are clustered into small pompoms that are covered in pollen. As the giraffe browses about the acacia tree, pollen sticks to its fur. When the giraffe moves to another tree, cross pollination may occur.

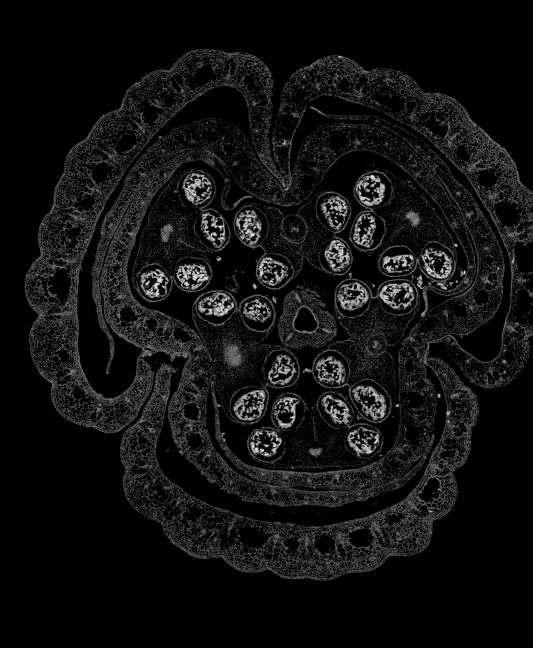

19 The identification of chemicals that stain different types of plant tissues was an important breakthrough in the interpretation of plant structure. This image shows a transverse section through the bud of a lily. The three sepals, three petals and six anthers are stained blue. This section is stained so that the developing pollen grains appear to glow inside the pollen sacs of the anthers.

THE MAKING OF SEEDS

By far the most significant of the domesticated polyploids is wheat. We owe our daily bread to a series of remarkable genetic modifications, which occurred over ten thousand years ago and was one of the forces that guided hunter gatherers into the role of farmers. Sometime, either before or during the earliest stages of the domestication of wheat, a wild, diploid species known as einkorn hybridized spontaneously with a diploid goat grass, whose identity is still a matter for speculation. In the course of this hybridization the chromosome complement spontaneously doubled to produce tetraploid wheat, known as emmer, which subsequently became one of the most widely semi-cultivated grains. At a later stage, domesticated emmer hybridized with another goat grass and, by a process of simple addition, the chromosomes of the tetraploid emmer combined with those of the diploid goat grass. This produced the plant we know as bread wheat, which would become the mainstay of people living throughout temperate regions of the world.

Bread wheat is actually a hexaploid. Genes from three different species in two different genera contribute to its six sets of chromosomes. The fact that hybridization occurred is no more than might have been expected, and no doubt it occurred repeatedly, as it still does, wherever species of wheat and goat grasses grow together, as they do across large areas of the Middle East. The question we should be asking is what enabled the hybrids to persist and multiply until they became predominant in certain situations instead of being rapidly reabsorbed among the medley of wild grasses? Did human intervention have something to do with it or did it happen by chance?

Polyploids usually grow more vigorously than diploids and possess many desirable horticultural qualities. However, polyploids also display more genetic buffering than diploids. Genetic buffering refers to the likelihood that deleterious genes are revealed when plants are crossed. Self-fertilizing diploids, such as peas and barley, are weakly buffered since there are only two forms of each gene present and each of these is likely to be the same. Maize, onions and other cross-fertilizing diploids are more strongly buffered. There are still only two forms of each gene, but these might be slightly different to each other. However, in polyploids the number of forms of each gene depends on the number of sets of chromosomes present, and all of these may differ from each other. Polyploids are therefore likely to hold greater reserves of genetic diversity than diploids. The disadvantage of polyploids is that, compared with diploids and their simple, predictable Mendelian ratios, attempts to attain particular objectives are complicated. This is due to the difficulties of predicting the outcomes of any cross, and so of creating genetic combinations that produce significantly improved varieties.

This chapter has followed the process of making a seed, starting when pollen is transferred from the anthers of one flower to the stigmas of another and proceeding with the creation of an embryo when a nucleus from a pollen grain fuses with an egg cell within an ovule. What remains is for the embryo within the ovule to grow into a mature seed, ready to be launched into the world to make its own way. When all goes according to plan, the embryo grows within the ovule and storage compounds and other resources are moved to it from the parent plant. In due course the walls of the ovule harden to form the seed coat, and the tissues within it lose water and become desiccated. The seed is formed and is, technically, contained within a mature carpel, the fruit. Most farmers would be surprised to hear that they sow a field of wheat with fruits, not seeds. Gardeners would be equally surprised to hear that the lettuce seeds in the packets they buy are, technically, fruits. Every wheat 'seed' is composed of the tissues of an entire carpel, as is every lettuce 'seed'. Technical definitions can be very important – but in this book, as we go forward, if it looks like a seed, behaves like a seed and is treated as a seed, it will be referred to as a seed.

The proportion of its resources that a plant devotes to seed production, and the way in which it apportions those resources between the alternatives of producing a few large seeds or numerous small ones, is a fundamental strategic choice. It is key to the plant's survival, and is the subject of the next chapter.

CHAPTER FOUR

Strategies for Survival

The concept of a strategy appears at first sight an inappropriate one to use about seeds. Plants, after all, possess neither the foresight to plan nor the wit to trick. Yet effective strategies, amounting in fact to full-blown stratagems, are the means by which seeds survive and fulfil their destinies by producing seedlings. Each and every species has evolved its own, honed by the pitiless tests of survival or elimination, and tailored this to its circumstances, special needs and whatever particular qualities the species may possess. Neither foresight nor wit is required, simply selection. First and foremost, the seed has to make do with whatever resources it has inherited from its parents; these, inevitably, are a compromise, which ensures they are likely to fall short of the ideal. Seeds, after all, are costly to produce and filled with energy-rich storage compounds. Their production places a considerable strain on the plants that produce them.

The magic formula that every species must produce if it is to survive has two very simple terms – size and number. Large seeds provide a generous endowment to each offspring, investing the seedling with the wherewithal to compete successfully in the struggle to establish itself. Numerous seeds ensure the greatest number of opportunities to find suitable sites in which to grow. In the best of all possible worlds every plant would produce sufficient numbers of well-endowed seeds to combine both objectives – but resources are limited and compromises must be made between the ability to compete and the ability to find a site. Some species have chosen to take things to extremes. Orchids and broomrapes, for example, produce innumerable tiny seeds with derisory abilities to compete in order to maximize the possibility that some will find a place to grow. Oak trees, chestnuts and palms, on the other hand, produce large seeds. Their seedlings are formidable competitors, although the seeds themselves are almost immobile and depend on animals of various kinds for any hope of locating suitable places to grow. Most species occupy a medium ground, where compromise rules. Enough seeds are produced to satisfy the need to relocate, and each one contains sufficient storage compounds and other resources to fuel some prospect of survival in a competitive environment.

Most people have a vague impression that seeds in general are small objects, and small size is certainly part of the natural lot of a seed. But everything is relative – within an overall impression of smallness lies room for enormous variations in size between the seeds of one species and another. The difference between the smallest seeds, like those of orchids, and the largest, such as coconuts, compares with the ultimate extremes found in mammals – between a pigmy shrew and a blue whale. Within this range each and every species has its place, and a striking aspect of plant morphology is the narrow variation in size of the seeds produced by any particular species throughout its entire range. Such remarkable consistency provides powerful evidence of the critical relationship between the way resources are apportioned in the production of seeds and all the other elements that compose the nature and life cycle of the species.

Nevertheless there is still room for some variation, and the average size of seeds produced by different populations in different parts of its range will increase or decrease, depending on the circumstances in which seedlings are produced. In situations where suitable sites to establish are few and far between, large numbers of small seeds are likely to be produced. In places where numerous seedlings crowd together, vying for light, nutrients and water, fewer but larger seeds are the rule. Apart from selection for competitive edge or the ability to find a site, seeds vary in size depending on their position on a plant or within a capsule. Kernels of maize from the rounded tip of a cob, for example, are always smaller than those towards the base. Seeds produced in the first fruits of the spike of a foxglove are larger than those produced towards the end of the spike.

During the earliest stages of plant domestication, this may have provided the means by which the grain size of domesticated wheats, barleys, rye and other cereals increased. As settlements developed, the land around them underwent changes which favoured the establishment of cereals and other annuals. This reduced the significance of site selection, and at the same time increased the density of such plants – particularly during the critical phase of seedling establishment. Such a change would have favoured the selection of competitive seedlings – the product of grains with relatively generous storage reserves. Within the course of a few decades this would have been sufficient perhaps to double the average size of grains being harvested from semi-domesticated wheat, barley or rye.

Apart from getting the number/size equation right, every species has to devise a stratagem that ensures its seedlings emerge at a favourable time of the year. Seedlings which appear at the 'wrong' time of the year are destroyed by frost, drought or other climatic hazards, while those produced at the 'right' season have remote, but nevertheless crucially important, prospects of growing up, flowering and producing seeds.

Every species has a game plan based on the seasons when its seedlings emerge, and the proportions of them that forsake the relative security of the soil to commit themselves to germination. Survival of plants in the wild depends on getting the game plan right, but, as we shall see, the significance of these stratagems extends beyond their importance to the plants that devised them. They also fundamentally affected humankind's health, wealth and wellbeing.

Anyone might be forgiven for supposing a seed's essential function is to produce a seedling. Most of us deal with seeds as gardeners or farmers, and our expectations leave us with a simplistic view of what a seed should do. When we sow a hundred lettuce seeds and a hundred seedlings come up, we congratulate ourselves on our success. Yet such a complete and enthusiastic eruption of seedlings would be a suicidal strategy for most wildflowers to pursue. Within a day or two a large slug might come along and polish off every seedling! They might be shrivelled by drought or reduced to pulp by an unexpected frost – and with no reserve to fall back on, that would be that.

Seeds have a vital role in linking one generation to the next. Their function is to preserve that link by protecting the embryos within them from adverse conditions, producing seedlings when the time is right, and, more often than not, holding some in reserve in case of disaster. Very often there is no one best time to produce a seedling, but rather several promising seasons. These may vary from year to year and, as we shall see in the case of trees and other long-lived plants, years may pass before they occur at all. The test of a successful strategy is the survival of sufficient seedlings to replace their parents – not even generation by generation, but over a span of generations, within which the size of a population may fluctuate enormously. Purple foxgloves, for example, live in woodlands where opportunities vary from year to year as surrounding trees come and go in cycles that may spread over centuries. In sunlit glades they appear in thousands (20). As the trees grow and their shade increases, there may be only a foxglove here, another there. One plant can produce a hundred thousand seeds, and under favourable conditions they appear in their thousands; in unfavourable years not a single one may germinate. The population survives provided enough grow up, taking one year with another, to maintain the population; on average a single seedling from hundreds of thousands of seeds is sufficient. That sounds a less than demanding purpose in life.

But seeds and seedlings feed the world. By the time they have done so, and all those who depend on them for a living have taken their share, even such apparently overwhelmingly favourable odds do not guarantee survival. More often than not, by the time seeds mature on a plant, the world around has become an unfriendly place. It is quite unusual for freshly shed seeds to drop straight into situations favourable for germination and seedling emergence, and most have to wait until better times to return before they germinate. The bluebell is just such a plant (21). But bluebells in different countries may be very different things, and discovering what challenges plants have to contend with, and what part seeds play in meeting them, can be achieved only if we all agree on the identity of the plant we are talking about. The bluebell in question is the English bluebell, the Scottish hyacinth, the French *Jacinthe des Bois*, and presently, and hopefully finally, the plant known to botanists as *Hyacinthoides non-scripta*. From now on we will call it the bluebell.

The problems and opportunities that bluebells face can be appreciated better from a plant's-eye view than from our lofty perspective of the world in which they live. Much of the story of our association with seeds in this book is told through the actions or eyes of human beings. For once, it is time to allow a plant to be the centre of attention. To anthropomorphize about plants is a curious exercise, but an interesting one.

I am a creature of the woods. My world is dominated by trees immeasurably taller and more powerful than myself, against which I cannot compete, but whose protection and shelter soften the harshness of winter, temper the glare of the sun in summer, and discourage my more pushy neighbours on the forest floor. I exist because over the millennia my ancestors found ways to coordinate their growth cycles with those of the trees.

I emerge from the safe house of my bulb while the trees are still leafless in late winter, into a world swept by autumn winds and reduced by winter frosts. Before most of the plants in the forest are stirring, my leaves make the most of whatever sunshine the season provides in sheltered spaces under the trees. I produce my flowers in May, while there is still enough light and warmth on the forest floor for the bees to pollinate them. Then, as the leaves of the trees unfold above me, the world I live in becomes a cool, shadowed place scarcely reached by the sun. My flowers fade away; my seeds ripen and mature, and I return to my underground home.

I am a sociable plant, and prefer to live in dense colonies. My seeds are too heavy to be dispersed by wind, and I make no alliances with animals or insects to carry them far away. My flower stems collapse as the seeds ripen, spilling them on the ground in a ring around me, where my offspring will grow up as my neighbours. By now the overbearing shade of the trees and the ubiquitous presence of their roots below ground have turned my world into a hostile place, where my offspring must shelter in their seeds till conditions improve. Then, by following the strategy that is the inheritance of every bluebell, enough will grow up to sustain and reinforce the colony.

The strategy referred to by the bluebell, like that of almost all seeds, is based on a simple either/or binary response. In any situation seeds have two options: to germinate or to do nothing. Germination irrevocably launches the seedling into the world to make the best or worst of whatever conditions it encounters – a matter of life or death. Doing nothing keeps options open, enabling the seed to wait for changing seasons to bring other opportunities.

The world into which freshly shed bluebell seeds are cast is indeed a hostile one, although its mask may be falsely benign. Woodland soils in June can offer conditions not unlike those we provide when sowing seeds in a glasshouse in spring; sustainingly moist, encouragingly warm, comfortably shaded, and richly endowed with humus and other good things. Yet any seedling tempted to leave the security of the seed finds itself in competition with already well-established vegetation, accentuated by the heavy shade cast by the dense overhead canopy of leaves on the trees, and with almost certain prospects of drought to come when the soil dries later in the summer. Not surprisingly, bluebell lore instructs newly minted bluebell seeds to do nothing and, however enticing the conditions, their seedlings remain inside the seeds.

Months later, as summer gives way to autumn and the temperature of the soil falls, the seeds will stir into life. Tiny, ivory-white radicles, which eventually develop into roots, thrust their way from many of the seeds, and grow down into the layer of humus covering the forest floor. Falling leaves blanket the germinating seeds and shelter seedlings from the worst of the winter cold. By the time spring is in the air the roots are well into the upper layers of soil and already beginning to form tiny, nascent bulbs. Cylindrical, green cotyledons, or seed leaves, reach towards the light through the litter of fallen leaves. The seedlings are ready and waiting to take full advantage of whatever sunshine late winter and early spring may bring – before their parent bulbs have even started to produce shoots, and weeks before the buds on the trees break into leaf.

Bluebells steal a march on spring, and on many other plants, by germinating during the winter. They are ready to benefit from every hour of every day's increasing warmth and light by the time other seedlings are emerging. But their strategy does have an Achilles heel. The seedlings must survive the winter for it to work, and bluebell seedlings are not amongst the hardiest of plants. Prolonged frosts or other inclement conditions kill tiny seedlings, even beneath their blankets of leaves, and that is why wild bluebells grow only in the milder, oceanic parts of western Europe. As an insurance against catastrophe, part of each year's seed crop produces no seedlings in their first autumn. Some will do so the following autumn, while others, buried by the activity of worms, moles and other creatures, may become part of a reserve known as the soil seed bank. This is composed of seeds of many different kinds of plants, able to lie in the ground for years, decades and in some cases for centuries before they germinate. Foxglove seeds are eminently good at that, which explains the appearance of the plants in their millions when woods are felled, perhaps fifty or sixty years after foxgloves were last seen there.

The bluebell story describes a strategy whose success is amply and most beautifully demonstrated by carpets of bluebells beneath the trees of many British woodlands in springtime. But there is a twist in the tale, which could turn out to be highly significant if summers become warmer in the parts of the world where this plant grows naturally. Between a third and a half of bluebell seeds germinate after experiencing current average summer soil temperatures in southern England. Quite small increases in temperature could raise the proportion to more than 75 per cent; larger increases in summer temperatures might tempt almost 90 per cent of the seeds to produce seedlings. In other words, more seedlings would run the gauntlet of bad weather in the winter, and fewer would take the prudent option of remaining within the soil seed bank. Bluebells still grow in our woods because over the years they have achieved a balance between risk and security which, taking one winter with another, provides positive odds on survival. The riskier strategy of producing greater numbers of seedlings in the autumn and thus keeping fewer seeds in reserve could be disastrous – particularly in the plant's toehold in continental Europe, where it is precariously balanced on the edge of its natural range, and quite possibly across large parts of southern and eastern England too.

Bluebells appear to have little resilience in the shape of genetic diversity to fall back on, should changing conditions alter the world in which they live. But the diversity they need may exist in the form of a plant almost universally condemned and rejected by those concerned with the conservation of the

British flora. This is its close relative the Spanish bluebell, native to southern parts of Europe where summers are warmer than those in Britain.

Spanish bluebells are more robust than the native bluebell. They do better in gardens, where they tend to multiply rapidly, often to the point of being regarded as a nuisance, and during the latter part of the twentieth century were increasingly reported growing in the woodlands among native bluebells. When conservationists observed these incursions, their reflexes began to twitch. Their alarm increased when they realized that the two species hybridize freely and produce offspring with fertile seeds. Conservationists, to whom alien immigrants are anathema, reacted with shock and horror, but before joining the chorus we should ask whether these Spanish immigrants really threaten the native population. Might their genes be its salvation – at least in the warmer southern and eastern parts of its range – if summers become increasingly warm and dry during the coming years?

Bluebells provide a clear example of what a plant strategy is, how it works and the ways in which it affects a species' prospects of survival. Comparisons between the strategies of different species can reveal the affinities or origins of a plant, and may help us decide where the ancestors of cultivated plants once grew wild. Bluebells are so characteristic of British woodlands that it seems almost perverse to suggest the plant is neither a true woodlander nor even a northwestern European plant, but that is what its germination strategy tells us. The way in which the bluebell uses high soil temperatures during summer to prime seeds to produce seedlings as temperatures drop in late autumn and early winter is a strategy characteristic of Mediterranean annuals and bulbs, including daffodils, tulips and scillas. The bluebell is in fact a wanderer from further south. It has adapted to life in the woods by coordinating its cycles of growth and activity with seasons when the leaves are off the trees. By doing so it manages to avoid competition from the sun-loving grasses and broad-leaved herbs that would overwhelm it in the open.

Wood millet, by contrast, is a plant with cast-iron credentials as a true woodlander. This plant of moist, broad-leaved and mixed forests has an immense range from the British Isles to Japan, extending across parts of southern Canada and into the northeastern United States. Could a single strategy serve the needs of populations growing in such extraordinarily different parts of the world?

This is an obvious question to ask, but one with big implications. If the germination strategies of a species' seed changed as it extended its range, there would be no curb on a plant's ability to spread. If, however, populations retain very similar strategies in widely different parts of a species' range, it follows that germination strategy serves to define a plant's geographical range. Of course, it is not the only factor. A plant's germination strategy interacts with life cycle, drought tolerance, winter hardiness and other aspects of its biology.

Freshly shed wood millet seeds land in a highly competitive and dynamic environment. Overhead shade from trees, plus dense neighbouring vegetation growing in moist soils in partially sunlit, partially shaded clearings, not to mention constantly changing balances between sunshine and shadow, combine to offer seedlings only the remotest prospects of survival. Nevertheless, old plants die, new opportunities arise and, however unlikely the prospects may be, seeds are the only way in which losses can be replaced and new sites located.

Seeds that mature during the summer when the soil is warm may germinate within a few weeks – unlike bluebell seeds. Under favourable conditions perhaps a third will have germinated within a month or so, and as many as two-thirds may have produced seedlings by the time winter comes. The remaining seeds shut up shop for the winter. When temperatures start to rise in the spring, seedlings emerge from almost all of them – leaving a small residue to join the soil seed bank. This bare summary ignores subtleties of such seeds' germination strategy. In total it provides a finely tuned, impressively flexible response to the very diverse challenges to which plants are exposed in the dynamic, unpredictable environments typical of deciduous and mixed woodlands.

The process outlined above provides the basis for a one-size-fits-all germination strategy which is flexible to whatever situation wood millet finds itself in. Modifications in the numbers of seeds enable the plant to respond to different climatic conditions. In Norway and Scandinavia, the extreme north of its range, seeds do not mature until late summer or early autumn, when soil temperatures have started to fall and severe winter weather is imminent. None of the seeds has time to germinate until winter is over and temperatures start to rise again during the following spring, when seedlings appear in time to make the most of good growing conditions during the long days of summer. On the other hand, in northern Spain, the south of France, down the spine of Italy on the Apennine Mountains and in other southern parts of the wood millet's range, seeds mature early in the summer, with several months of high and increasing soil temperatures ahead of them. Provided the soil around them remains constantly moist, most of the seeds will germinate, developing into smaller plants long before the onset of winter. The seedlings reap the benefits the following spring, when they are big enough to hold their own in the hurly-burly of renewed life and activity that makes this such a competitive season for plants.

Scandinavia and southern Europe are the extremes of wood millet's range, but in Britain and across northern France and Germany its seeds mature around midsummer. There is time for some seeds to germinate before falling soil temperatures call a halt, but many do not do so, preferring to wait till the following spring to produce seedlings. The proportion of seeds that germinate depends on their location and the conditions they encounter. High proportions of seeds that land up in sunlit, sheltered sites produce seedlings, especially in warm years. Seeds in more densely shaded situations, especially in cool years, produce fewer seedlings during the summer; most seeds wait till spring. A simple, basic strategy enables seeds of wood millet populations, throughout its whole range, either to produce seedlings during the summer or to do nothing till spring – in accordance with the opportunities or hazards posed by the local environment.

In dealing with seeds, we are almost invariably interested in obtaining seedlings – and we pay little attention to those that do not germinate. Under natural conditions, however, the latter are no less crucial to the survival of the population than those that germinate, and when we think about them the idea of a straightforward binary response – offering the alternatives of germinating or doing nothing – begins to look a little too simplistic. Seeds that do not germinate but wait for another opportunity are not inert. They are aware of, and able to respond to, conditions around them – particularly temperature – by adjusting their metabolism in preparation for what the future brings. Bluebell seeds, for example, respond to warm soils during the summer by priming themselves to produce seedlings when temperatures fall in autumn. Cold weather during the winter enables wood millet seeds that did not germinate during the summer to produce seedlings as soon as temperatures start to rise in spring. Gardeners have known for a long time that seeds of plants from cold regions, notably trees and shrubs and alpines, will often not germinate until they have experienced several weeks, or even months, of exposure to temperatures just above freezing point. This simple and effective strategy prevents seeds from germinating late in the summer or early autumn, even though the soil may be enticingly warm and moist, only for their seedlings to be destroyed during the winter ahead.

Gardeners sow seeds to produce plants for their vegetable or flower gardens. Scientists study the conditions under which seeds germinate. The expectations of both tend to induce a positive view of germination, making it hard to come to terms with the propensity of seeds to lie apparently inert for months or years. When gardeners sow seeds and no seedlings appear, they call it a 'failure', and account for the non-appearance of seedlings by describing the seeds as 'dormant'. They talk about 'conditioning' seeds by chilling or other treatments to 'break' dormancy and 'make' them produce seedlings. The jargon is so well-worn that no one notices that it reveals nothing about what goes on inside the seeds themselves. Nor does it acknowledge the essential part played by plant stratagems in timing germination, nor the vital contribution to the long-term survival of populations made by seeds lying in the soil seed bank.

Gardeners use the word dormant as a convenient figure of speech to describe seeds which do not produce seedlings when expected to do so. Scientists use the word as though it describes a phenomenon – one without shape or form, limits or boundaries, since nobody has satisfactorily defined it. It is as though we used the word sleep to indicate indiscriminately the state of insensibility, whether that was slumbering, lying in a coma, stupefied by drink or concussed from a blow to the head. Vagueness and imprecision of that sort would make it impossible to progress in the study of sleep or what part it plays in our lives, yet research on dormancy in seeds has been based on such woolly ambiguities for decades. The concept of dormancy reveals a great deal about our presumption of control over the seeds we sow. It masks an ignorance so profound that it has seriously impeded progress in understanding seed physiology for the last half century.

Our concept of dormancy has led to the presumption that 'dormant' seeds must somehow differ from those that produce seedlings, albeit in unspecified ways, defined only by the fact they do not germinate when their fellows do. Sometimes differences can be attributed to obvious conditions, such as the possession of impermeable seed coats, but whether a seed germinates or does nothing appears more often to be entirely arbitrary. In response to a flash of light some seeds produce seedlings; others do nothing.

In the absence of any discernible physical differences between the seeds that germinate and those that do not, thoughts turn to explanations based on the existence of a genetic difference – a 'dormancy factor' serving to inhibit or promote germination. Such an explanation would imply that seeds in the soil seed bank are genetically different from the rest of the population. However, this would make little sense in the natural context of the survival of a population of plants at a particular location.

Genetic differences are all too readily invoked to explain any differences between one plant and another. But even plants (and animals) with similar genetic constitutions do not necessarily all respond to events in the same way. What we observe depends on how the genotype interacts with the environment, and what we see, known as the phenotype, is the visible expression of that interaction. Thus, the number of seeds that produce seedlings in response to particular temperatures or a flash of light depends on how their genes make them react to the conditions they encounter. The reaction may be an all-or-nothing response, as it would be if dormant seeds were genetically different from those that produce seedlings.

Interactions between genotype and the environment may be expressed quantitatively in the proportion of seeds that germinate; in other words, what we see happening expresses the probability that a seed with a particular genotype will produce a seedling in a particular situation. When seventy out of one hundred seeds in a germination test produce seedlings, we should not jump to the conclusion that the other thirty are different in some way.

The result may simply express the probability that each seed has a 70 per cent chance of producing a seedling. The acid test, were it possible, would be to do the germination test, then pack the seedlings back into their seeds and repeat the test with the same batch of seeds. Seventy per cent of them would produce seedlings, as before. If this was a phenotypic response they would not be the same individuals, but a random assortment of those that had made the cut. If genetic differences accounted for those that produced seedlings and those that did nothing, we would expect to find exactly the same seeds germinating the second time around.

Phenotypic responses can be elusive, which makes those that are easily recognized all the more welcome. One such is the stratagem adopted by wood millet seeds. We might explain the observation that high proportions of seeds from the southern parts of the plant's range germinate in the summer, and only a few produce seedlings the following spring, while northern plants produce few or no seedlings in the autumn and a great many in spring by supposing that few seeds produced by southern populations are dormant, in contrast to those produced by northern populations. This leads logically to the conclusion that selection had fine-tuned the proportions of dormant and non-dormant seeds in different populations to match the con-ditions in different parts of the distribution of wood millet – in other words, that seeds from the north were genetically different from those in the south.

However, experiments show that wood millet seeds from all these populations respond in the same way to a variety of standardized conditions. This makes it highly unlikely that they are genetically different, or that particular seeds are dormant and others non-dormant. Populations growing in different places all possessed the same germination stratagem, and the differences in the proportions of seeds that germinated and those that did nothing arose from the ways in which this stratagem interacted with climatic conditions – specifically soil temperature. At least as far as wood millet is concerned, this tells us that germination strategies do not change as a species extends its range – but rather that species' ranges expand within geographical/climatic limits defined by the stratagem.

On first impression, the strategies adopted by the seeds of different plants appear to multiply endlessly. However, plants that grow naturally under similar conditions tend to share broadly similar strategies, making it possible to recognize them as a Mediterranean Strategy, a Steppe or Prairie Strategy, a Deciduous Woodland Strategy and an Alpine Strategy. Gardeners can afford to regard such variations as challenges that add interest to the craft of gardening, but farmers operate on a scale which makes it impracticable to tailor operations separately to the needs of each and every species.

Farmers can only grow plants that meet the demands of agriculture, and one of the most powerful qualifications is the possession of a germination strategy amenable to cultivation. That still remains the *sine qua non*, without which no crop can be cultivated on a field-scale.

When a farmer sows a field, he does so in the expectation that almost every seed will germinate within a short time, before weeds have gained a head start, to produce plants of even age and size. Those who farm in temperate, rather than sub-tropical, parts of the world expect seedlings to be produced even at the low soil temperatures typical of early spring. Seeds with special needs, for example a predilection for germinating at particular times of the year or a prudent inclination to set seeds aside in the soil seed bank, create insuperable problems for anyone wishing to grow a crop, harvest it and the next year perhaps sow a different crop in the same field, without the complications caused by the germination of seeds left over from previous years. Seeds that behave like those of bluebells or wood millets are thus not amenable to large-scale cultivation.

This means that in our search for clues about where plants were first domesticated, we should look for parts of the world where the wildflowers' germination strategies meet farmers' specification for a satisfactory crop plant. This is essentially one in which all seeds germinate, and the seedlings are produced rapidly and at low temperatures. The first place to look is among the annual flowers of the world. Almost all temperate cultivated crops, and most vegetables, are either annuals or biennials. However, annuals in general are scarcely more likely to be endowed with amenable germination strategies than perennials. Biologists who study annual weeds, for example, are often frustrated by their reluctance to produce seedlings to order. Seedlings of groundsel, fat hen, purple dead nettle, poppies or annual mercury appear spontaneously in their thousands in fields and

20 The purple spikes of the foxglove are found by their thousands in open areas; in shaded areas only a few plants may be found. However, each plant may produce hundreds of thousands of seeds, only a few of which need to survive to produce the next generation of plants.

21 English bluebells on the floor of deciduous woodlands are one of the enduring images of spring in the United Kingdom. Bluebells flower before the woodland floor is too shaded by trees and before they can be outcompeted by other plants. One of the ways they achieve this is by having seeds that germinate during the winter.

22 The Soviet geneticist Nikolai Vavilov (1887–1943) was one of the pioneers of understanding the importance of genetic variation in the conservation of crops. Vavilov's work in the early twentieth century was the inspiration for the foundation of numerous seed banks in the latter part of the century.

23 Vavilov and his collaborators identified eight centres of crop origin or crop diversity.
(**1**) China
(**2**) Southeast Asia and the Pacific Islands
(**3**) Central Asia
(**4**) Near East
(**5**) Mediterranean
(**6**) Ethiopia
(**7**) Mesoamerica
(**8**) South America and Andes

gardens, but when their seeds are collected and then sown the results are unpredictable and usually disappointing.

Annual plants survive by seeking gaps, as and when they occur, in situations that are more or less permanently occupied by perennial plants. In the wild such opportunities occur sporadically, maybe at long intervals. The soil seed bank is thus an essential element in their strategy as a place where their seeds can wait, perhaps for decades, for the chance to grow. Cultivated ground provides seeds with more or less perpetual opportunities and plenty never encountered naturally, and they make the most of it. But prudence is maintained even in the midst of plenty and high proportions of each year's seed crop do nothing rather than germinate, adding to the credit balance in the soil seed bank. The abundant seedlings produced are the product of many different years, not the spontaneous, simultaneous emergence of those most recently shed; hence the old saying 'one year's seeding; seven year's weeding'.

Annuals, in general, may fall short of the farmers' expectations. But when we focus on those that grow naturally around the Mediterranean basin and adjacent parts of the Middle East, we find a more promising situation. The Mediterranean climate is generally consistent with strongly marked characteristics, of a kind which might be expected to give rise to distinctive germination strategies. Summers are hot, dry and inimical to plant growth of any kind without supplementary water. Winter is the season when rain falls and temperatures, though cool, are seldom lethal. The strongly defined climate provides a clear-cut optimum season for the emergence of seedlings, with little room for variations, and autumn comes with seeds of thousands of annual wildflowers lying in the soil ready to go. They germinate with the first heavy rains of autumn – the sooner the better, in order to keep ahead in the race to establish a place in the world. Seeds grow through the winter, often at low temperatures, sometimes in poor light, making the most of days when the sun shines and the air warms, and by spring are ready to burst into flower. For a month or so the land is filled with flowers, then summer drought sets in; flowers fade, seeds mature, are shed and scattered on the ground where they lie through the hot dry summer waiting to renew the cycle the following autumn.

One among the many flowers that make southern Spain and the Balearic Islands so colourful in early spring is the rosy catchfly. Its seeds are easy to collect, and if we came across it on holiday we may be tempted to pick a few ripe capsules in the hope of growing the plant in our gardens. If we sowed the seeds as soon as we got home, however, we would probably be disappointed. Annuals that germinate immediately before the start of hot, dry summers do not survive around the Mediterranean. The catchfly survives because its freshly shed seeds will germinate only at temperatures well below those of the soil in early summer. As the seeds lie in the soil they undergo internal changes, known as after-ripening. These progressively broaden the range of temperatures at which they will germinate, greatly speed up the time they take to do so, and reduce the numbers that do not germinate. By autumn virtually every seed is just waiting to produce a seedling as soon as the soil cools and rain falls.

The rosy catchfly's strategy is just what farmers look for in a crop plant. More or less simultaneous ripening of seed at the conclusion of the growing season facilitates harvesting. The need for a few weeks of after-ripening before seeds become capable of producing seedlings prevents premature germination, particularly important in stopping grains of wheat or barley germinating before they can be harvested in wet years. The temporary nature of the condition means that, after being stored for a short time, almost every seed germinates within a few days at temperatures likely to be experienced by autumn-sown crops, or produces seedlings remarkably rapidly at the much lower temperatures encountered by crops sown early in spring. In the case of catchflies, pleasantly attractive to gardeners, but of no interest whatever as crop plants, all this might seem a trifle academic. However, similar germination strategies are shared by wildflowers and annual grasses all around the Mediterranean, and by many which grow far to the east across the foothills of the mountains of Syria, Iraq and southern Turkey. Throughout this area numerous annual wildflowers share germination strategies of exactly the kind that farmers look for in cultivated crops. It thus comes as no surprise to discover that so many of the crop plants, vegetables and flowers grown on farms and in gardens originally grew wild in these parts of the world.

The germination strategies of the wildflowers matched the requirements of cultivation so closely that after domestication they remained unchanged. Lettuces, for example, and corncockles both have a history of cultivation extending several thousand years – one in the vegetable garden, the other as a cornfield weed – and, despite all the selective pressures they have encountered along the way, they possess germination strategies similar in all respects to the catchfly's. Indeed, corncockle seeds growing as cornfield weeds in Britain germinate in precisely the same way as those from plants growing wild in their original home. One of the most valuable features of the Mediterranean germination strategy lies in the seeds' abilities to germinate at low temperatures, enabling them to perform well when sown into still cold soil in early spring. Some seeds, lettuces for example, are so tolerant of cold that they produce seedlings even when lying on blocks of ice.

Until less than a century ago our ideas about the origins of wheat, barley, lentils, onions, lettuces, cabbages and other ancient, long established crops were based on excavations and discoveries in the classical sites of antiquity around the Mediterranean and eastern Asia. These crops were cultivated in ancient Egypt, as well as Mesopotamia, Asia Minor and the Indus Valley, yet archaeological and linguistic evidence revealed they had originated in none of them; their places of origin, some thousands of years earlier, appeared to have been lost in the mists of time. Maize, French and runner beans, peppers, potatoes and tomatoes were recognized as sixteenth-century introductions to the Old World from the New World. They were known to have been cultivated in ancient times in Central America and along the Andean spine of South America, but where their wild ancestors had grown was a matter of guesswork. The discovery of the places where the wild ancestors of these crops had once grown owed less to archaeological excavations than to the efforts of plant collectors based in the Soviet Union or the United States of America. Between the 1920s and 1960s they set out to make comprehensive collections throughout the world of different varieties of crop plants, in support of national plant breeding programmes.

The most notable of these collectors was a Soviet scientist called Nikolai Vavilov (22), a major player in the events described in Chapter 7. In the course of their travels Vavilov and his colleagues made hundreds of thousands of collections of economically important plants. As his collections accumulated, Vavilov began to notice patterns that had not been perceived before. Others had worked within limited areas, and their perceptions had been bounded by the horizons of that area. Vavilov jumped the boundaries, his view of the whole world giving him new perspectives. His first discovery was that crops were far more diverse than anyone had previously realized. His second discovery was that the greatest diversity was concentrated in a few, clearly defined parts of the world. And, most surprisingly, those regions were more often than not in remote places (23).

Vavilov's interest had been attracted in the first place by the diversity of the strains of particular crops (landraces) grown by peasant farmers in the Soviet Union. Now he could see that differences between, for example, landraces of Soviet wheat, though very significant to plant breeders, were small compared to the differences found across the plant's entire range. Across the length and breadth of Europe and North Africa, and east across the southern Soviet Union, Vavilov found that variations in wheat were centred on a few themes. When he looked south to Anatolia, Transcaucasia and the foothills of the mountains fringing northern Syria, Palestine, Iraq, Iran and Afghanistan, the number of themes multiplied enormously, as did the number of variations within them.

Vavilov found similar patterns in barley, rye and lentils, and in his collections of flaxes, vines, pistachios, melons, figs and numerous other secondary crops. Every village seemed to have its own particular, often totally different, kinds. He had set off presuming he would find the greatest variety in the more sophisticated parts of the world, where farming methods were advanced and the plant breeders had been at work, and had expected to find fewer and less interesting different kinds in less developed places. Yet the pyramid had been turned on its head, with a broad foundation of diversity firmly based in parts of the world where agricultural techniques were least developed.

Eventually Vavilov concluded that differences must accumulate over the years, and the places where crops were most varied were where they had been cultivated longest – and this, he argued, pinpointed their wild origins. He coined the term 'centre of origin' to express this idea, and accordingly proposed that the Middle East, from Anatolia to Iran, was the centre of origin of wheat – and for good measure some eighty-three other crops, including barley, rye, oats, chick peas and lentils.

Vavilov extended the idea to other parts of the world, where crops were notably varied, and came up with another eight centres of origin. To the west of the Middle East, contiguous with it and merging in places, was the area around the eastern Mediterranean. Here he found exceptionally numerous varieties of lettuces, peas, radishes, cabbages, broad beans and other vegetables, as well as lentils, olives and vines. Whereas the Middle Eastern centre had provided for the farmers, the Mediterranean centre was a treasure hoard for gardeners.

Further to the east lay the Central Asian centre in the mountains of Tajikistan, Turkmenistan and Uzbekistan, where numerous kinds of wheat, rye, lentils, chick peas and other pulses supplemented those already found elsewhere, accompanied by apricots, apples, melons and other fruits, as well as carrots and several other vegetables. Ethiopia was identified as a centre for wheat, barley, chick peas, beans and lentils of various kinds. However, since no wild relatives of these plants were known from the region, Vavilov concluded that their diversity had sprung from plants brought into the country in ancient times by migrants or traders.

A separate and completely distinct range of crops, including rice, several kinds of millet, buckwheat, soybean and a variety of other beans, as well as numerous vegetables quite unlike those grown in Europe, was found during expeditions to China. These established the credentials of China as one of the world's most significant centres of domestication in antiquity. Collections from India provided evidence of the vital role played by its inhabitants in the introduction of tropical crops and vegetables, including rice (one of several places where this crop appears to have been domesticated), and lentils again, as well as cucumbers, aubergines, mangos and other tropical fruits, and a number of spices. The list was reinforced by prolific collections of sugarcane, different sorts of rice, and yet more tropical fruits and spices in southeast Asia and Malaysia.

The New World displayed a similar pattern of clearly defined localities, containing an almost bewildering complexity of crop varieties, contrasted with far more extensive regions where relatively few different kinds were found, most of which had probably been imported from elsewhere. One of the most productive parts of the world was in the mountainous regions of Central America, notably in Mexico, where villagers in every settlement and valley cultivated distinctively different sorts of maize, tomatoes, upland cotton, beans, squashes, chillies and sweet peppers, as well as numerous other crops and vegetables. A similar story with regard to potatoes, sweet potatoes, tomatoes, tobacco, pawpaws, quinoa and peppers emerged along the spine of the Andes through Bolivia, Peru and Ecuador, and on Chiloe Island, off the coast of southern Chile, Vavilov came across a remarkable variety of different kinds of potato. Finally he identified southern Brazil and parts of Paraguay as a centre of origin for manihot, peanuts, chilli peppers and pineapples.

Eighty years have passed since Vavilov pinpointed his centres of origin – eighty extremely active years during which our knowledge of the world and its plants has increased at an accelerating rate. By and large, however, his concept has stood the test of time well. No further centres have been proposed, though a good case could be made for Western Europe as the centre from which grasses, clovers and other fodder plants were distributed around the world during the colonization, in recent times, of North America, New Zealand and, to a lesser extent, Australia and South Africa. Further experience suggests that the Chinese centre would be better split into several centres, including the north for millets, several pulses and numerous vegetables of the cabbage family, the south for rice and soybeans, and the southwest for buckwheat and, probably, barley. Vavilov also stretched the argument a little too far in his use of the phrase 'centre of origin', since areas of high diversity may be the product of factors other than crop origin, as is the case with Ethiopia. The term 'centre of diversity' provides a more neutral, perhaps more academically sound, expression of the concept, but leaves the question of origin open.

The fact remains that almost the entire gene pool of cereals, vegetables and other plants on which mankind depends for survival was generated in a few, limited regions of the world. Consequently, the genetic diversity from which crop improvements will be made in the future resides within the plants growing in these same restricted regions. During the last fifty years each and every one of these places has been subjected to severe changes, depleting the genetic resources they contain and now threatening to destroy them entirely. Their destruction would wipe out a huge part of the genetic base on which plant breeders still depend for the improvement of crop plants and vegetables.

The obvious question posed by Vavilov's centres of origin is why are they so few and so clearly defined? Hunter gatherers roamed across the surface of the globe in all directions. They gathered and ate plants across the entire land mass of Eurasia, yet the secrets of domestication were revealed only to those living in circumscribed areas along parts of its southern fringes. People also roamed the whole of North America, but, apart from exceptions such as the sunflower, the North American wild rice, crook-neck and other squashes and Jerusalem artichokes, gathered by the indigenous population but probably domesticated by Europeans, very few of the many plants gathered across the whole extent of what is now the United States and Canada achieved any significance as cultivated plants. The same pattern applies to huge expanses of Africa and South America, yet all are places where, once the secrets of domestication had been revealed elsewhere, the practice of cultivating crops spread to every corner, wherever climatic conditions allowed.

We also need to ask to why certain parts of the world provided us with so much, whereas other places with similar climates gave us little or nothing. The clans of the San people hunted game and gathered the roots and fruits of wild plants along the Atlantic seaboard of South Africa, Aborigines lived off the land across the length and breadth of Western Australia, and Native Americans in California and indigenous people in central Chile also lived where climates were essentially similar to the Mediterranean. Yet no vestige of agriculture developed in any of these places, and none of the plants they gathered became domesticated. The hunter gatherers remained hunter gatherers, even though at least some of the plants they used for food would have been amenable to cultivation. Evidently the customs, culture and tribal structures of the indigenous people were essential components of the equation too.

The most skilful cultivator can achieve nothing until seeds produce seedlings, and seeds fail to do this for numerous reasons. Some deteriorate within weeks of being harvested; they die long before the time comes to sow them. Others germinate only when they encounter particular conditions; without those conditions, no seedlings appear. Many seeds germinate sporadically; a few seedlings emerge one week, a few more the following month, some not till a year, even two years later. However desirable the product may be, these never produce sufficient plants at a particular time to provide a worthwhile crop. Seeds of some species are tardy risers, who lie in the soil for weeks after they have been sown; weeds have a head start and the young crop may be smothered at birth. A wildflower has no future as a cultivated plant unless it produces seeds that store well, that germinate soon after they are sown, and that do so in sufficient numbers to produce the dense, uniform number of plants needed to produce a crop. It makes things easier, too, when seeds are neither so small that they are lost in the soil as soon as they are sown, nor so large that they are lost to birds and rodents.

Other regions, apart from the Mediterranean, provide conditions that lead to the development of germination strategies amenable to cultivation. The climate in mountainous parts of Mexico, for example, is the antithesis of conditions around the Mediterranean. Dry, chilly winters are unsuitable for plant growth, and instead the growing season extends from spring through warm, wet summers till autumn. Yet these mountain valleys are the cradle of cultivated forms of sweet peppers, tomatoes, beans and maize. Though so different, however, the climates of the Mediterranean and Central Mexico share a crucially important feature. Transitions from favourable to unfavourable seasons are clear-cut and consistent from year to year. Both contain a distinctive time to grow, a time to produce flowers, a time to ripen seeds and a time to produce seedlings – and in both, when the time comes to produce seedlings, the victors are those that germinate without delay.

Nevertheless, the fundamental difference between these two climates is that while the Mediterranean has a winter growing season, Mexico has a summer one. Crops whose ancestors once grew around the Mediterranean can be sown early in the year, while soils are still cold after the winter, but they tend to cease growth and run to seed in the summer. Tomatoes, maize and peppers, by contrast, succeed only when their seeds are sown after the soil has warmed, and revel in whatever heat the summer brings.

The criteria that define whether a wildflower's germination strategy makes it amenable to cultivation have had far-reaching consequences. Germination stratagems determined appropriate species for crops and the places where hunter gatherer societies could develop into settled farming communities. In most parts of the world the local inhabitants had no chance of ever becoming farmers because few, if any, of the plants they gathered for food were amenable to cultivation. The wildflowers whose seeds, tubers and fruits fed them were not inadequate or insufficient as sources of food. These plants never became cultivated plants because they simply did not possess the germination stratagems that made it possible to cultivate them.

CHAPTER FIVE

Travellers in Time and Space

'Where are we?' I looked at Chris. He shook his head and with a shrug of the shoulders raised a hand, palm upwards, into the air. It was a question we asked each other several times a day, and one that once was on the lips of plant collectors everywhere. Chris Humphries, then a student at Reading University, later a leading light in the Department of Botany at the Natural History Museum in London, was my companion in Greece on my first seed collecting expedition in 1970 on behalf of the Seed Bank at Kew. We were part of a team working through the former Yugoslavia, Greece and Bulgaria, and the plant group legumes was a focus of our collecting.

We had maps of two kinds – German road maps, telling us exactly where we were as long as we stayed on a road, and British military topographical maps, first prepared in the 1930s and barely revised since, displayed forbidding, but authentic-looking arrays of mountains, gorges, valleys and rivers. Each time we made a collection, we had to record its geographical location and height above sea level. That would be no problem today, with satellites overhead and geopositioning devices in hand. But before the days of satellites our ideas about which point on the map we had reached and how high we were tended to be fairly hazy after several hours spent traipsing around mountains. We entered brave guesses, like many plant collectors before us, with a general expectation of getting the country right, and, with a bit of luck and inspiration, being within a few miles of our guesstimate.

If Chris and I were often not sure of our whereabouts, how did the seeds we were hunting know where they had reached? Any living organism responds to the world around it, but it can only do this in what might be described as a logical and sensible manner if it is able to establish its whereabouts in a way appropriate to its needs. Plants live in a world almost beyond our imaginings. Their lives and responses are incomprehensible in the familiar terms by which we assess the impact of situations and events, and establish our place in the world. When the sun shines we seek shade, when hungry or thirsty we look for food or water, when hot we find ways to cool down. Plants respond to sun and shade and drought, but cannot react to alleviate the stresses imposed by such things.

As mammals our body temperatures remain constant, however warm the day or cold the night. A seed, in contrast, warms with the day and cools with the night, its temperature keeping step with nature. By responding intimately to every variation, seeds are able to use temperature as an indicator of what is going on in ways denied to us. As we saw in the previous chapter, they can use it to respond to passing events and prepare for the future.

The stratagems by which seeds ensure they produce seedlings at appropriate seasons provide broadly based game plans. They are not concerned with details, and although they may clear a seed for germination, the local situation may be such that it would be most unwise for seedlings to emerge. Seeds may be so deeply buried in the soil, for example, that no seedling could struggle to the surface. Seedlings might be smothered by established plants before they could find a roothold and space to grow. Acting blindly is not a recipe for survival. Seeds need to be able to perceive what is above them, to feel what is going on around them and to locate their orientation and position, wherever they may be. Those might seem impossible attainments for objects that possess neither eyes nor other identifiable senses.

Seeds, if we could read the information they receive, and find ways to tailor them to survive the ordeal and pass on their experiences, would make ideal, small, light, ready-made time/space capsules. We could send them off to outer planets to transmit information about the conditions they encounter. Seeds could gauge and record temperatures, mark temperature differentials between night and day and measure the length of the days, the intensity of sunlight, the presence of overshadowing obstructions, the status of soil water, the passage of time, the season of the year, the oxygen tensions in surrounding soil and the concentrations of carbon dioxide and ethylene. They do all these things naturally, hour by hour, day after day.

Before getting carried away by visions of planetary, even intergalactic, space/time capsules, let us see how seeds move around planet earth. This brings us back to the recurring theme of plant immobility. The popular conception of seeds is that they enable plants to move to new places. Generally speaking this is accurate, as witnessed by the extraordinary variety of devices designed to give seeds mobility or induce some passing animal to provide them with a lift. Nevertheless, plants can, and do, use seeds to consolidate and enforce possession rather than seek out new places to grow. Bluebell seeds, for example, fall to the ground around the parent to form small groups, which year by year expand into the large, densely packed, spectacular colonies.

Annuals around the Mediterranean and in neighbouring parts of the Middle East also use their seeds to reinforce territorial possession. The sun burns off much of the vegetation during the summer, leaving more or less extensive areas of bare ground ready for colonization when cooler, moister conditions return in autumn. Such spaces are an open invitation to invasion by annual species (p. 98). Once installed, they use their seeds to reserve space for continued occupation in future years. Where their parents lived and died, they, too, will live and die, and if conditions favour them, and they can hold their ground against invasion by shrubs, they may come up year after year in the same place – turning annual occupation into perennial possession. Under natural conditions the annual lifestyle is a precarious one, but when nomadic hunter gatherers became increasingly sedentary, they changed the balance of existence of local annual plants. Gathering firewood, grazing and folding sheep and goats, constructing settlements and moving around them, served first to create, then to preserve and enlarge patches of bare ground open to colonization by annuals. Year by year, as seeds of one generation after another fell to the ground, wheats, goat grasses, barleys and similarly endowed annuals grew ever more thickly as they consolidated their hold on the spaces made available for their occupation.

Nevertheless, the seeds of most species are adapted to get away from their parents, and the ways in which this is done are legion. Many enlist the wind (24). Ash, maple, birch, hornbeam, as well as pines, firs and many other trees, produce seeds with wings effective over a few hundred metres at most. Dandelions and thistles equip each small seed with a filamentous parachute which a brisk gale can carry for kilometres, although most travel only modest distances. Willows and poplars produce masses of tiny seeds caught up in a network of the finest and lightest filaments that drift with the lightest winds. Relations between parent and offspring reach a low point with the orchid; they sacrifice almost everything to distance one from the other, including the seed's very prospects of survival. Orchid seeds consist of a cluster of undifferentiated cells virtually bereft of food reserves, enveloped in a fragile net so buoyant that it lifts into the air at the slightest breath. Orchid seeds can drift for hundreds of miles, crossing continents and seas with few prospects, or none, of coming to rest anywhere they will survive. Even if the seeds are lucky enough to find themselves in a promising situation, their lack of any resources to support growth leaves them dependent on finding a fungal companion to nourish and support them. Why seeds of plants whose best prospects of finding suitable places to grow lie close to home should commit themselves to almost inevitably futile long-distance travel remains a mystery. Many species of orchids have to produce hundreds of millions of seeds in the course of their lives, as a consequence of this seemingly perverse lengthening of the odds against survival.

Some seeds are seafarers. Like the coconut, the characteristic palm of tropical beaches, they drift with the currents that cross the oceans, sometimes travelling for thousands of miles before being cast up on a shore (25). Most other seeds are killed by immersion in seawater, but a few – and particularly several members of the pea family, in which the embryos are protected by impervious seed-coats – can float with the currents for months, even years. Not surprisingly, plants which grow naturally among islands are most likely to be adapted to saltwater transport, and the Caribbean islands and the Indonesian archipelago are the homes of some of the best known. Seeds from the Caribbean travelling by Gulf Stream and the North Atlantic Drift find their way often enough to the western coasts of Britain. Here they have generated a small folklore about the mystery of their origins, acquiring names such as sea hearts, sea purse and sea coconut, which reflect their links with the oceans.

Animals, too, in all shapes and forms, have been enlisted to remove seeds from the vicinity of their parent. Some do this passively. New Zealand burr weeds, herb Robert and goosegrass – or sweethearts – equip their seeds with hooks, grapnels and other devices which attach themselves to the fur of passing animals and the socks of hikers for free rides to somewhere else. Seeds also get caught up in the wool of sheep. In the early twentieth century, lands around the Yorkshire woollen mills, where wool from overseas was dumped or stored, were once happy hunting grounds for botanists on the look-out for exotic plants. Eventually so many 'shoddy' plants were found that a book, *Alien Plants of the British Isles* (1994), was compiled.

Podocarps, the great southern hemisphere branch of the conifers, went a step further, along with their cousins, the yews; they resorted to bribery. Instead of waiting passively for animals to pick up their seeds as they passed by, they packaged them in sweet, brightly coloured, fleshy wrappings (known as arils) or enclosed them in succulent, plum-like fruits to attract birds, bats, tree-living mammals and anything else in search of a feast. The seeds of cycads, immovable in anything less than a Force 10 gale, are also wrapped in sweet, fleshy coverings. Animals remove the large seeds of cycads from their cones, nibble their sweet coatings and eventually deposit them. If the cycad seed is lucky, it will be in a place where it can germinate and grow.

The appearance of fruits marked the start of a long process of alliances between plants and animals, and was another major milestone in the evolution of life on Earth. The ploy was successful – animals ate the fruit and deposited the seeds away from their parent trees. Nowhere was this strategy more successful than in New Zealand. As the islands rose out of the sea about twenty-five million years ago, they were reached only by birds and a few reptiles. No mammals and no snakes colonized them before they were cut off from the rest of the world, and so New Zealand became the kingdom of the birds. They dwelt in luxuriant evergreen, temperate forests, enjoying a benign climate and few predators. Many abandoned flight altogether in favour of less energetic ways of moving around. Such congenial conditions created a unique situation, however. In the absence of serious predation, overpopulation became a greater threat to the survival of bird species than the need to maintain positive balances between prey and predator, which drives reproduction elsewhere in the world. The birds had to find ways to produce fewer offspring, or they would eat themselves out of house and home.

The sugary offerings of the podocarps were the answer. They supplied ample calories and enough nutrients and vitamins to support this undemanding lifestyle, but insufficient protein and other essentials to produce more than a bare minimum of eggs. Small clutches, laid infrequently, became the rule. Kiwis laid and incubated a single huge egg; other bird species nested every other year or so. The arrival of the Maoris was to change all that. They ate the birds, relishing to the point of extinction the moas, adzebills, flightless geese, swans and other large poultry. Then Europeans came, and proceeded to cut down the forests and introduce rats, cats, ferrets and possums, thus decimating the smaller species. Today conservationists struggle with the problem of persuading what is left of the New Zealand birds to produce at least a clutch of eggs a year.

New Zealand was the only place in the world where plants could get away with feeding animals almost exclusively on jelly babies in exchange for distributing their seeds. Elsewhere, while some jelly babies were most acceptable, they had to be backed up by more substantial fare, and in tropical forests the figs stepped into the breach. An astonishing eight hundred and fifty different species of fig grow in one part of the world or another, and are generally left alone by locals and loggers because they provide neither timber nor useful materials for building houses or other purposes. Yet figs are among the most valuable, wildlife-friendly trees, full of nooks and crannies where birds can nest and animals seek shelter. They also produce enormous crops of succulent fruits on which monkeys, parrots, lemurs and toucans feed. Every day of the year, somewhere in the forest, a fig tree will be bearing ripe fruit, and wherever it may be, birds will fly and mammals beat a path to it.

Succulent fruits of many kinds supplement the basic diet of figs in the tropics, and are sought after by animals in colder parts of the world as well. But hips, haws and other fruits with a modicum of starch and tiny quantities of protein, sweetened by large amounts of sugar, are not enough to counter winter frosts and periods of snow when food of any kind is in short supply and survival depends on fat reserves built up from more generous fare. Fieldfares, redwings and waxwings relish fruits as a dessert, but supplement them with insects, worms and other sources of protein. Plants, too, rose to the challenge of providing more sustaining fare, but in doing so made a sacrifice which fundamentally changed the balance of advantage between them and the animals that distributed their seeds.

Previously, the bait had been the succulent tissues of the fruits, not the seeds themselves. The latter were eaten when the fruits were eaten, but most were protected by hard, indigestible seed coats. They passed through the bird or animal undamaged, to be deposited somewhere far away from their parents. Now the substance of the seeds themselves would be the currency in which animals were to be paid, not sugary attachments or pulpy wrappings. The concept of obtaining services in exchange for cheap food, originally devised by the podocarps, had grown into a much more expensive and demanding operation. It was now one that involved the loss of a proportion – often a very high proportion – of the plants' offspring.

Hazels, walnuts, chestnuts and oaks maintained the long-established symbiotic associations with animals by producing large seeds packed with carbohydrates, oils, fats and other foodstuffs, highly attractive to hungry birds and mammals. The effort is justified provided that just one acorn, on average, out of the enormous number produced by an oak, is carried away and buried in a place where it produces a seedling, grows into a tree and in due course produces more acorns. All the rest are sustenance for vast numbers of birds, mammals, insects and fungi, most of which play no part in furthering the prospects of oak tree survival. Benefits to the tree appear to be minimal, but we assume this expenditure of resources really is necessary – presumably because opportunities for seedlings to establish occur so infrequently that seeds have to be ready and waiting to make a break at any time.

Plants which produce more orthodox, small, dry seeds meet this situation by salting seeds away in the soil seed bank where they can lie for decades, even centuries, waiting for the chance to germinate. But opening a seed bank account is not an option for oaks and nut trees. Their large, moist seeds cannot survive in the state of suspended animation so readily achieved by small, dry seeds, and, even if they could, are far too attractive to creatures of too many kinds to be left lying around on the ground waiting to be incorporated into the soil seed bank. So oaks – and some nuts too – have developed partnerships with squirrels, jays and other animals that carry away fallen acorns and bury them in the surrounding countryside. These partnerships even extend to matching the size of the acorns produced by the predominant kinds of oak trees in different parts of the world with the carrying capacity of the local birds' beaks – small acorns for small birds, larger acorns for bigger ones. Such arrangements tuck the acorns safely away out of sight of predators, but, sadly for the acorns' prospects of survival, jays and squirrels have proved phenomenally good at remembering where they have buried their trophies. The acorns' best chances of survival lie in the premature decease of the animal before it comes back to recover them.

The oak tree is a conspicuous example of the apparently profligate way in which seeds provide for the world around them. However, it is only unusual because acorns are large objects that reveal the extent of the situation particularly clearly. A simple way for plants to cut back on these free handouts would be to produce poisonous seeds; they are extremely well-equipped to produce poisons, and such an enterprise would present few problems. Nevertheless, the benefits of enlisting animals to spread their seeds around the world evidently outweigh the costs of feeding them, since only a tiny minority of plants produce poisonous seeds. Acorns illustrate the point. Their only hope of a future lies in finding some way to remove themselves from the shadow of their parents, and since they are too heavy to drift in the wind, an animal porter is the most feasible alternative. Seedlings growing up in the shadow of their parents have little hope of survival. Apart from the shade, and the remoteness of the possibility that the tree will keel over and let in the light, seedlings receive a rain of mildew spores and leaf-eating caterpillars from overhead, which overwhelm them sooner rather than later.

A few plants have, as noted, gone down the poison path – notably leguminous shrubs. Legumes have partially resolved the problem of removing their seeds from the parents' vicinity by producing mechanical devices which propel seeds ballistically for several metres once the seed pods mature. Distribution in time (i.e. seeds from different generations germinating at different times and thus exposing different gene combinations to selection) was achieved by enclosing embryos in impervious seed coats. Having sorted out those problems, the plants had less need of the distribution services of birds and other animals, and some of them discouraged seed predators by packing their seeds with poisonous alkaloids. However, a group of small beetles, the bruchids (familiar to farmers as pests of beans), developed immunity to the poisons. By stuffing themselves with poisonous seeds, bruchid grubs become poisonous themselves, and immune to the attentions of most predators.

Once an acorn, or any other seed, finds itself in the soil in reasonably clement conditions, one might suppose its troubles would be largely over – apart from the prospect that any seedling it produces will most probably be eaten or overwhelmed by competing vegetation. However, this is far from the truth.

To land safely, especially if you are a tree, can be the prelude to an immensely long future. However, there is no guarantee of a future or even of immediate survival, and, because trees are so long-lived, it is more than likely to be a hopeless venture in at least ninety-nine years out of a hundred. A small grove of large trees in the shadow of Mount Rainier in the coniferous forests of America's Pacific northwest sums up, and places in stark perspective, the hopelessness of a seed's prospects. The Grove of the Patriarchs, on an island in the Ohanapecosh river, is a tight little community of Douglas firs, western red cedars and western hemlocks, some of them more than a thousand years old. This grove is sometimes said to be one of the finest pieces of old growth forest in the northwest. This is a surprising thing to be told about a group of trees in which decrepitude is more apparent than thrusting vigour. Some are semi-decayed with large limbs dropping off, others have already fallen and their huge trunks lie rotting slowly on the ground. Those that remain in reasonably good health often grow higgledy piggledy. The most decrepit trees are the Douglas firs, the more presentable the western hemlocks – not because they are longer lived than the firs, but because they represent the final phase in the forest succession, the heirs of the estate originally colonized by the firs.

Centuries ago a forest fire prepared the way for the Douglas firs. Their seedlings sprang up soon afterwards from seeds blown in from occasional survivors and neighbouring trees. They grew in an almost pure forest, not unlike a forestry plantation, for some hundreds of years. As the shade from the fir foliage increased and layers of humus built up on the ground from fallen needles, seedlings of western hemlock and red cedar grew up from airborne seeds which blew in between the trees from parents in established patches of nearby forest. Meanwhile, year by year, seeds produced in the cones of the Douglas firs matured and drifted down through the trees to the shaded, softly receptive floor of the forest below. However, for century after century, even as their parents began to collapse and fall to the ground and the hemlocks and cedars took over the forest, quite possibly not one Douglas fir seed produced a seedling. The conditions were not right for germination. One day another forest fire will sweep through the Grove of the Patriarchs, killing the cedars and hemlocks, burning away the humus-rich organic covering on the forest floor and providing the mineral substrate and light which Douglas fir seeds need in order to produce seedlings.

The seeds of many trees, conifers in particular, display preferences for particular conditions which – in nature and forestry practice too – determine whether seedlings live or die. Douglas firs are a pioneer species that colonize ground after the passing of forest fires, hurricanes, volcanic eruptions, landslides and other destructive events. Their seeds germinate on bare, exposed, mineral soils, often low in humus. They grow up in each other's company, without the support and shelter of established trees. Giant redwoods are pioneers too, as are junipers, cypresses and the New Zealand kauris, as well as Monterey, lodgepole and ponderosa pines. Western hemlocks, on the other hand, are settlers, which prefer established conditions. Their seeds germinate on the humus-rich forest floor. Their seedlings grow up in the shade and shelter of existing trees, waiting centuries if need be for opportunities to take over from the pioneers that prepared the way for them. Coast redwoods are also settlers, as are many firs, false cypresses, spruces, yews and some pines.

The monumental timescale of a forest means that centuries may pass during which prevailing conditions make it virtually impossible for seeds of particular species to produce seedlings. Seeds may survive in the soil seed bank for a few years, a few decades, but their life spans are scarcely significant. The tree itself confers the longevity that makes the seeds they produce travellers in time, ready when opportunities arise to produce the next generation of seedlings.

Lesser plants seldom match the timescales to which trees live, but they can still spring surprises. Surveys of populations of herbaceous perennials in meadows have revealed that some may live for decades. The youngest plants in a group of Pasque flowers growing on downland in southern England, for example, may be thirty or forty years old. Seeds are produced year after year, but the advent of a new plant in the colony is a rare event. This phenomenon justifies the belief of conservationists that moderate numbers of seed can be gathered, even from rare plants, without jeopardizing the species' chances of survival.

Many members of the pea family have developed a simple but effective way to prolong the life of their seeds. The seed coats become impregnated as the seeds mature with layers of fat and chitinous substances, making them impervious to water and gases from the outside world. The embryos within them can survive almost indefinitely, isolated from the world, but, with no water or oxygen to sustain them, they are incapable of producing seedlings. Months, maybe years, later, wind-blown particles of sand, nibbling insects or invasion by fungi wear away or reduce the integrity of the seed coat. Water and gases reach the dormant embryos within, which awaken and produce seedlings. It is a somewhat haphazard system, but effective as a means of spreading germination out over many years.

There are records of seeds still remaining capable of producing seedlings after spending years in the cupboards of museums and herbaria. In September 1940 two incendiaries and a large oil bomb smashed through the roof of the Botany Department of the British Museum in London. Edgar Dandy, the botanist on fire duty that night, promptly put the fires out, saving the collections but soaking many of the books and specimens in the department. Some seeds of the Chinese silk tree, collected during the Macartney mission to establish diplomatic relations with the emperor of China in 1793, received a cold douche, and seized the opportunity, after nearly a hundred and fifty years, to germinate and produce seedlings. In 2006, there was short-lived public interest in the seedlings that the Royal Botanic Gardens, Kew, had managed to germinate from three seeds discovered in the British National Archives. The seeds, *Liparia viposa* and a protea and an acacia, had been collected by a Dutch merchant, Jan Teerlink, during a trip to the Cape of

Good Hope in 1803, and, for seeds, had been stored under less than ideal conditions. More dramatically still, a group of researchers was able to germinate a seed of the sacred lotus from a Chinese lakebed. The seed was carbon dated at 1,288 years old.

Low temperatures and low water content provide the recipe for the long-term survival of seeds. That covers pretty well everything except for the large, obviously moist seeds of nuts and similar objects such as acorns, and a miscellaneous collection which includes the seeds of citrus trees, the handkerchief tree, some maples and the large seeds of many tropical trees, among them cocoa and rubber, which are intolerant of desiccation. Seeds of these species are short-lived. They die if they lose too much water, or if temperatures fall below freezing point and ice crystals, formed inside their cells, damage cells' membranes and walls.

During the 1930s and 1940s, experiments on seed storage in the USA and elsewhere showed that seeds could survive at temperatures far below zero, when dried to moisture contents below five per cent. For practical purposes, commercial cold stores and domestic deep freezes provide appropriate temperatures, and widely available commercial desiccants such as silica gel dry the seeds effectively. These findings led to the installation in 1958 of cold stores for the long-term conservation of the seeds of crop plants at the USDA's National Seed Storage Laboratory at Fort Collins in Colorado. This organization would subsequently provide the science behind the use of seed stores for the worldwide conservation of crop genetic resources.

So far in this chapter, seeds have played the part of pawns blown by the wind or carted around by animals. They are helpless, cast by fate into situations where they have no hope of producing seedlings and transported by forces over which they have no control into places they cannot choose. After seeds have been carted, deposited or blown to whatever destinations they finally land up in, they cease to be passive, however, becoming instead masters of their own destiny – or, more correctly, the destiny of the seedlings they will produce. The sensitivity with which they gauge their surroundings, assess events and future prospects accurately, and germinate when the moment is right is critical – a matter of life and death to their seedlings.

Germination is an event, not a process. Once the trigger has been pulled, there is no going back. From seed to seedling is a matter of a moment, though it may be days before the evidence for the change in the form of seedlings actually appears. Stratagems prepare seeds for germination, and until whatever conditions they prescribe have been satisfied, seedlings will not be produced. Once they are satisfied, seeds are primed ready to go. Provided they are adequately supplied with water, oxygen and other basic necessities, all they need is the right stimulus to set them off. Once launched in this way, metabolic activity starts to accelerate. Storage reserves are mobilized, embryos activated, cell growth initiated and, within days, roots and shoots push their way into the world.

The stimuli that pull the trigger are sunlight and temperature, independently or in combination. We can all appreciate how seeds sense temperature; even inanimate objects become warmer when temperatures rise and cooler as they fall, and the metabolisms of all living organisms are directly affected by temperature. But light is something different. Animals see light with their eyes – not necessarily complex organs like our own, but at least identifiable light receptors – and respond to it through their brains. Seeds patently possess neither eyes nor brains. However, the first step towards sensing light is to absorb it, and anything coloured (and that simply means anything that is not white), absorbs light. The brownish, blackish surfaces possessed by so many seeds are highly light absorbent, but that does not mean they are capable of 'seeing' light. The essential requirements for sensing light are the possession of some means of harnessing its energy to change or accelerate some aspect of metabolism, and a means of transmitting that changed state or accelerated reaction to a receptor of some kind. Their view of the world may be different from ours, but seeds can see.

Photosynthesis, the fundamental life-support system of plants, and ultimately of animals, harnesses the energy of the sun to produce sugars; it is the most familiar way in which plants use light. However, photosynthesis is just one system among several in which light is used to drive or control growth and development. Chlorophyll is the green pigment that gives leaves their characteristic colour and captures energy from sunlight during photosynthesis. Various other pigments contained in plant cells drive other processes, among them the one that enables seeds to detect and respond to light.

How do plants respond to the world in which they grow? That is one of those naive questions that can lead to remarkably profound and far-reaching answers, but until less than a century ago botanists were led by its very naivety to accept answers that begged, rather than answered, the question. Orthodoxy declared, in rather vague terms, that plants grew and ordered their lives by responding appropriately to soil, climate, changes in temperature and daily light levels. This mishmash of sometimes conflicting, changeable and frequently aberrant stimuli was credited with providing plants with the information they needed to produce flowers in season, shut down in advance of winter and go dormant before cold weather set in, start growth again in spring (all too frequently in 'unseasonably' cold conditions) and perform a hundred other complex organizational activities. Many or most of these not only coped with events of the moment, but also managed to anticipate events as well.

In 1918 two plant physiologists, Wightman Garner and Harry Allard, working at the USDA's research station at Beltsville, just outside Washington DC, became interested in the strange behaviour of two cultivated plants. One was a variety of tobacco called Maryland Mammoth. This plant was a boon to Virginia tobacco growers because it seldom, if ever, flowered. It just grew and grew, and so long as it grew it continued to produce leaves, to the profit of the tobacco growers. Another plant that obstinately marched to its own tune was the Biloxi variety of soybean. This had everything to commend it apart from its insistence on producing beans at a time of its own choosing. The growers sowed seed at intervals of two or three weeks to spread the harvest, only to find that – irrespective of when they were sown – every bean matured at the same time.

Garner and Allard grew seedlings of the tobacco and the soybean in pots, and left half to grow outside all summer long. They moved the others into a dark shed every afternoon, bringing them out again each morning. Plants of Maryland Mammoth given the shed treatment produced flowers long before the summer was out, while those left outside just kept on growing. Soybeans moved into the shed flowered five weeks earlier than those left continuously out of doors.

Garner and Allard must have had something in mind when they spent a summer taking plants in and out of a shed, and that was most probably a suspicion, converted into a hypothesis, that the time when Maryland Mammoth and Biloxi soybean plants produced flowers had something to do with the length of the day. The shed experiment provided a simple, expedient way to test that conviction, but it did not prove that day length had anything to do with what happened. Plants in a shed experience different temperatures to those left outside. Changes in humidity are dissimilar to those in the open air, and their daily ration of sunlight is greatly curtailed.

Any of these, and other more subtle factors, might have been responsible for the results. However, later experiments proved that the length of the day was indeed the key to the production of flowers. Day length was also the key to unlocking the secrets of how plants tailor all manner of other responses to their environment. Leaves fall in autumn as day length declines. Annuals around the Mediterranean produce flowers in response to increasing day lengths in early spring, in order to ensure that their seeds ripen before the onset of summer. After a spring spurt, the growth of many perennials ceases and flowers appear as days lengthen. Other plants, including tobacco and soybean, take full advantage of what summer has to offer, and growth gives way to flower production only as day length declines with autumn's approach. The leaves of many spring-flowering bulbs die and the plants retreat underground in response to lengthening days, even in wet seasons which might tempt them to remain longer above ground. Every one of these is a forewarned, forearmed response in which what is happening in the present is used to prepare for events to come.

Garner and Allard coined the terms 'photoperiod' to describe day length and 'photoperiodism' to describe the phenomenon by which plants respond to the length of the day. As a foundation on which to base responses, it has the enormous merit, compared to other climatic features, of consistency. The length of the day in every part of the world follows a regular annual progression, predictable and constant from year to year. At its simplest, this means that a plant able to measure the length of the day can confidently look forward to the onset of summer at the coming of the spring equinox, and prepare for winter when the lengths of the night and the day are again equal in the autumn.

In order to be able to respond to day length, plants must be able to 'see' light and, more intriguingly, have some way of measuring its duration. Years passed before it was discovered how this was done. The major contributions, fittingly, were made by physiologists working at Beltsville in the laboratory where Garner and Allard had made their original discoveries. Between the 1940s and 1960s Sterling Hendricks and Harry Borthwick gradually narrowed the search, starting with the discovery that light at the red end of the spectrum had very strong, positive effects on seed germination and flower initiation. They then claimed to have identified a pigment as the light receptor involved in photoperiodic responses and, with the aid of a biophysicist called Warren Butler, named it phytochrome. Unfortunately, it was present at such low levels that it defied all attempts to extract it, provoking sardonic comments about their invisible pigment from rivals who doubted its existence. Eventually, and doubtless to his enormous satisfaction, Hendricks presented the world with a small phial containing a suitably exotic-looking turquoise liquid. He was not only able to convince the doubters it was phytochrome, but also to provide a satisfactory explanation of how it worked, and an analysis of its chemical constitution.

Phytochrome is a protein which absorbs light at the red end of the visible spectrum, close to the limit of human perception. Its particular claim to distinction lies in the fact that it is what is known as a photochromic pigment, which means that it changes when illuminated. As phytochrome absorbs light, it turns into a different form. The two forms of phytochrome are recognizable by their slightly different colours, which are detected by the wavelength of light, measured in nanometres (nm), that they absorb. One form, the one in Hendricks' vial, absorbs most light in the red part of the spectrum (650–670 nm), while the other form absorbs wavelengths just beyond the limit of vision for the human eye, at the far-red end of the spectrum (705–740 nm).

This Jekyll and Hyde protein enables plants to see light, to use the information they receive to assess situations, and to modify their behaviour accordingly. Red-absorbing phytochrome firstly provides plants with a simple and highly effective way to establish the presence and direction of a light source. Secondly, because after passing through a leaf the spectral composition of light shifts towards the far-red end of the spectrum, it enables plants to detect the presence of overhanging foliage. Thirdly, because far-red phytochrome gradually reverts to red phytochrome in the dark, it provides plants with the means of measuring the length of the night. The longer the period of darkness, the more complete the reversion. In fact, despite the emphasis on photoperiod and the length of the day, plants actually respond to the length of the night. However, the conventions are well established and well understood and, attractive as splitting hairs about whether we are talking about short day or long night may be, plant scientists have sensibly decided to let them stand.

Phytochrome regulates growth and development in many ways, most relevantly the germination of seeds and the growth of seedlings. Because light does not penetrate far into the soil, it enables seeds to assess their depth of burial and avoid germinating when they lie so deep that their seedlings would be unable to reach the surface. The effect is reinforced by registering changes in temperature as night follows day, since these are greatest on the surface and tail off to more or less constant temperatures a short way underground.

Overhanging foliage is another hazard for seedlings. We observe shade as a reduction in overall light intensity. Seeds register a crucial shift in the spectral composition of the light because red light is absorbed while passing through a leaf; the consequent increase in the proportion of far-red light provides a warning of the presence of other plants. This alerts seeds to the presence of established competition which might smother newly emerged seedlings. If seeds ignore the warning signs and produce seedlings despite foliage overhead, the seedlings' phytochrome kicks in once more to control their growth, enabling them to grow rapidly towards wherever the shade is least dense.

Strangely, although seeds use phytochrome to see and respond to light, there are virtually no reports of day length controlling germination. The strategy responsible for the success of Mediterranean annuals, for example, is based on responses to temperature that avoid germination during the heat and drought of summer and promote the rapid appearance of masses of seedlings as temperatures fall in the autumn. The same result would be achieved by producing seeds which germinated in response to sequences of declining day lengths, corresponding to those experienced as days shorten in the autumn. There are many other situations in which seeds respond to temperature, where strategies based on day length appear likely to be at least as effective. The reason may be a simple practical one. Responses to day length are typically, perhaps exclusively, mediated through leaves that are fully exposed to light, whereas seeds often find themselves in situations where the length of the day could not be accurately measured.

For us, the choice was a fortuitous one. If Mediterranean annuals had opted to respond to day length, rather than temperature, many of the vegetables and flowers which are the mainstay of our gardens would be of little use to gardeners in all but the warmer parts of Europe and North America. There would be no point sowing seeds of popular vegetables, including lettuces, broad beans, spinach and peas, in the spring, nor those of well-loved flowers such as larkspurs, marigolds, cornflowers and love-in-the-mist, because they would lie in the soil without producing seedlings. Only seeds sown in the autumn would germinate, and these plants could never have been successfully grown in gardens where winters are too cold for their seedlings to survive. Similarly if photoperiod, not temperature, controlled the time when wheat, barley and other major crop plants from the Middle East produced seedlings, we would not have had the option of sowing them in the spring. These crops too would have been confined to places where winters are not too severe for their seedlings to survive. Without the option of spring sowing, which until quite recently was the norm, it is doubtful indeed whether arable farming or effective vegetable gardening would have spread far beyond its cradle.

Laboratory experiments have shown that, unlike responses to photo-period, which by definition depend on long periods of illumination, seeds can react to light almost literally in a flash. Very short exposures, measured in seconds, can provide the stimulus needed to produce seedlings, and observations suggest that even the brief turning over of the soil during ploughing can be sufficient to trigger germination of many arable and garden weeds.

Lots of attempts have been made to prepare lists of seeds that need light in order to produce seedlings; of those that germinate better in perpetual darkness, and of others that are not affected much one way or the other. The lists are more remarkable for their lack of consensus than their usefulness, particularly those that purport to tell us which seeds germinate best in total darkness. Germination in total darkness is a troubling specification, since seeds appear to have much to gain by using light as a guide and much to lose by launching their seedlings into a world of darkness. Misleading results may be produced by the lamps used to provide light in laboratory experiments. These are not the equivalent of sunlight, and they emit wavelengths that can inhibit germination. Temperature can also strongly influence responses to light, and those used for routine tests in laboratories may not reflect natural conditions. Tests tend to be done at 20°C or more – temperatures towards the upper limit of, or considerably above, those at which seeds of plants in temperate parts of the world germinate under natural conditions. Responses to temperature and light often go hand in hand – seeds may use both to detect how deep they lie in the soil – and it is quite normal to find responses to light varying with temperature. Responses to light and dark can be satisfactorily defined only by looking at how seeds behave over a wide range of temperatures, including the daily ups and downs as night follows day. To study the effects of temperature on germination accurately, an experimental set-up is needed that allows precise control of temperature.

Within a few months of my arrival in 1964 at the Jodrell Laboratory at the Royal Botanic Gardens, Kew, seed germination became firmly established as the main research topic of the new Physiology Department. During the first few years, responses to light appeared likely to be more promising and rewarding than those of temperature; reflecting the widely accepted orthodoxy. In 1966 I went to Beltsville to learn more about studying the ways in which seeds respond to light. However, Borthwick and Hendricks (whom I met) were both close to retirement, and I became fascinated by the problem of temperature. I was introduced to equipment that would transform my work upon returning to Kew – a thermo-gradient plate constructed by Martin Jensen, who was studying the effects of temperature on seed germination. He had attached heaters along one edge of a square aluminium plate and, by cooling the opposite edge with a refrigerant, produced a continuous gradient in temperature from one side to the other. Seeds sown on moist filter paper, laid on the aluminium, experienced temperatures ranging from *c.* 35°C to just above freezing point depending on their position.

Back at Kew, David Fox, who managed the workshop, produced a thermo-gradient plate of his own. Instead of a plate he used an aluminium bar with precisely controlled arrangements to heat one end and cool the other. Sets of bars provided more flexible experimental set-ups than a single plate and, after a little trial and error, continuous temperature gradients from 0°C at one end to 40°C at the other could be maintained. These thermo-gradient bars quickly became the main experimental facility used in our experiments.

Germination response to temperature could be studied in greater detail than anyone had dreamed of doing previously. Among these experiments were investigations of the responses of different vegetables to temperature. We were astonished to discover that seeds of extremely ancient crops, including lettuces, onions, leeks and tomatoes, germinated under precisely the same conditions as their wild ancestors would have done, thousands of generations ago. There was no reason at all to believe that their germination stratagems had been much, if at all, affected by cultivation. We collected seeds from populations of wildflowers growing in different parts of their range and compared the conditions in which they produced seedlings. More often than not, the places where seeds had been collected had little or no effect on their germination. Irrespective of where they grew, within a species, the seeds of most populations, or the individuals that occur in a specific area, possessed similar strategies – though different species, like different vegetables, varied widely in the actual strategies they displayed.

Temperature provides the foundation on which other seed responses are based. Reactions to light modify responses in vitally important ways only after exposure to sequences of temperature have brought seeds to the point of being capable of germinating, and only when temperatures are appropriate for the production of seedlings by that particular species. After dispersal from the parent plant, seeds first have to interact with whatever environment they land up in. Some germinate immediately, whatever the conditions, including the seeds of belladonna lilies, nerines and crinums, which more often than not start to produce at least roots and often shoots even before they drop. The seeds of some mangroves not only germinate, but even grow into fair-sized plants before they fall from their parents to embed themselves directly in the mud below, or drift away to be cast by currents on some distant shore.

The seeds of the majority of species tune in to what is going on around them, responding to temperatures and changes in temperature. They lie for weeks, months or sometimes years in the soil, exposed to the changing temperatures as one season follows another, making internal adjustments to their metabolisms which later facilitate the production of seedlings. Eventually they encounter conditions that tell them the time has come to germinate.

Seeds occupy a unique position in the life history of plants, and there is one further feature of their construction that makes them unusual. They are chimaeras – organisms composed of two or more distinctively different sets of tissues. Inside is the embryo – a new entity with a new genotype, formed by the putting together of chromosomes from both the male and the female parents. Outside is the seed coat, a tissue derived from the ovule wall and made entirely of cells with the genotype of the mother plant. Single-seeded fruits, those of lettuces, strawberries and nuts, for instance, might be described as more of the same thing; their embryos are enclosed by cells derived from the carpels of the mother plant in addition to the ovules, so they might be said to consist of a double dose of mother. But the mystery deepens. The seeds of flowering plants contain a third, genetically distinct tissue – the endosperm. This is formed when a second nucleus from the pollen cell enters the ovule and fuses with two nuclei formed during the cell divisions that produce the egg cell. The result is a two-parts female, one-part male triploid tissue, which nourishes the embryo during its early development and, in some species, contains a major part of the seed's storage reserves. In the grasses, such as wheat, rice and maize, it is endosperm that feeds the human race.

The next chapter looks at ways in which gardeners regard seeds and how they use them to produce the plants they want to grow. Their encounters with seeds frequently lead to frustrating situations when the gardener's desire to produce seedlings does not correspond with the seeds' readiness to germinate. In such situations, an understanding of the whys and wherefores underlying seeds' behaviour crucially affects the chances of persuading them to produce seedlings.

CHAPTER SIX

Seeds in the Garden

Farmers are in thrall to seeds; gardeners aspire to master them. This was not always so, and is only partly true today. The roots of western European gardening, like farming, lie in the Middle East and around the Mediterranean, where many plants produce seeds with germination strategies that make them naturally amenable to cultivation. Gardeners benefited from this, just as farmers did, and a high proportion of the vegetables grown in kitchen gardens are descended from plants that 'volunteered' for domestication because their seeds germinated rapidly, completely and at low temperatures. At one time the ancestors of a long list of vegetables – including lettuces, endives and chicories, spinach and fat hen, onions, garlic, leeks, shallots and other alliums, peas, lentils, chickpeas and many beans, turnips and rape, carrots, parsnip and skirret, scorzonera and salsify, radishes, fennel and alexanders, as well as numerous herbs with aromatic leaves or seeds and opium poppies – were among the hangers-on, or commensals, in and around fields of primitive cereals. Neither encouraged nor discouraged, these were plants whose natural attributes fitted them to the demands and needs of cultivation, with qualities that made them worth gathering and using in one way or another.

In time, as such useful plants became increasingly appreciated and regularly used, they were sought out and gathered, to play greater or lesser parts in the daily diet. Later their seeds would be gathered and deliberately sown to perpetuate them. The veggie patch was born, later to come of age as the kitchen garden. Although the seeds of different vegetables germinated under conditions that made them amenable to cultivation, they did not share identical responses, nor have they adapted to cultivation by adopting uniform responses. After thousands of generations in cultivation, seeds of lettuces, leeks, onions, celery, radishes, carrots and cabbages retain the germination strategies of their wild ancestors.

Books about farming and gardening, when they mention the subject at all, usually contrast the complexities of the germination responses of wild-flowers with the absence of restrictions characteristic of, and indeed essential for, cultivated plants. These books frequently assert the belief that seeds of cultivated plants have adapted to cultivation by losing the inhibitions characteristic of wildflowers. But evidence does not support such a presumption. First, the seeds of long established cultivated plants germinate just as their wild ancestors did, and second, there is a tendency to exaggerate problems associated with the seeds of wildflowers. Many do indeed cause problems, as we will see. However, wildflowers in certain parts of the world are the natural possessors of responses which make them ideally suited to germination. That does not apply only to those species selected by time and circumstances for cultivation. A survey done at Kew showed that seeds of more than 50 per cent of over six hundred species tested produced high proportions of seedlings within an acceptable time. Undoubtedly, botanic gardens tend to grow species amenable to cultivation, for the very good reason that otherwise they might not be able to grow them at all. Yet, however much this tendency is discounted, the results still show that a great many species among representatives of the sixty different families included in the survey possessed at least the basic germination responses sought in a cultivated plant.

The discovery of the New World by Europeans in the late fifteenth and sixteenth centuries led to the introduction of new vegetables and other plants, including several among the most valued of those we grow today. Potatoes, maize and tomatoes were far and away the most significant. Potatoes are the only major crop plant not grown from seed in temperate parts of the world. Others, which play greater or lesser parts in gardens depending on location, include sweet and chilli peppers, quinoa and amaranth grains, pumpkins and a remarkable array of squashes (26), sweet potatoes and kumara, manioc, peanuts, French, scarlet runner, Lima and other beans, Cape gooseberries and tomatillos, as well as tobacco and numerous fruits of one kind or another. All originated in mountainous parts of tropical regions, predominantly in central Mexico and along the Andean spine of South America, and they needed to be treated accordingly. Plants from these parts of the world grow during the summer in luxuriant conditions, quite unlike the comparative austerity to which ancestors of the pre-Columbian European garden flora were adapted. European gardeners, brought up to be wary of over-indulging Mediterranean plants, inclined to grow fat and lazy when spoilt, discovered that their subtropical American cousins respond to luxury and do not thrive without it.

Seeds of these newcomers possessed two of the three qualities for which gardeners hope. High proportions germinate, and do so rapidly to produce even stands of flourishing seedlings, but, loyal to their origins, they do not germinate at low temperatures. Thoughts of early crops from early sowings before the soil starts to warm up – a process to which species from the Mediterranean are so well adapted – had to be put aside. New ways had to be found to grow vegetables which only flourished if they were not expected to tolerate adverse conditions to anything like the extent that gardeners steeped in age-old methods expected of their plants.

Not surprisingly, the runner and French beans, maize, tomatoes, peppers, squashes and other introductions from the New World quickly became established in southern Europe, where they grew well in the long hot summers, especially when irrigated. However, these plants struggled in the cooler summers of the north, where the frost-free growing period was shorter, and for many years played little or no part in gardens in cooler parts of the continent. Among other reasons, the suspicion that they might be poisonous weighed heavily. Tomatoes, potatoes and capsicums, all members of the nightshade family, were all deemed too close for comfort to some of the most notoriously deadly poisonous plants known. Although grown as a curiosity by gardeners in Paris and London by the end of the sixteenth century, the tomato was not widely adopted, even in the gardens of the wealthy, until well into the eighteenth century. Even a plant as useful and comparatively easily grown as the potato, destined to become the staple diet of millions of people, was adopted cautiously and by slow degrees.

In *The Natural History of Selborne*, published in 1788, Gilbert White remarks that 'potatoes have prevailed in this little district by means of premiums within these twenty years only; and are much esteemed here now by the poor, who would scarce have ventured to taste them in the last reign'.

Lettuces from the Mediterranean, maize from central Mexico and dozens of vegetables, cereals and other crop plants with similarly amenable germination strategies provide an aspect of gardening which rewards those who take pride in attention to detail and are most confident when following long-established routines. Provided the rules are followed, a select list of plants can be relied on to provide high yields and top quality. No alternatives to wheat and barley, maize, rice or potatoes produce equivalent crops, nor are there substitutes for most of the major vegetables.

Gardening in the flower garden may have quite another face, one where impressions and ambience count for more than maximum productivity. Here diversity, novelty and flexibility are valued more than uniformity and predictability. Such a dichotomy between productivity and beauty is probably as old as cultivation itself. The earliest farmers, boasting around their hearths about whose were the most promising, the best grown, the most vigorous or productive patches of einkorn and barley, contrasted with those early gardeners who cherished the wildflowers around their huts, encouraged their presence and took pleasure in their beauty.

Some gardeners like to follow as well as impose rules, and are more comfortable with agricultural crops rather than venturing into unfamiliar places and delighting in the rewards they bring. Other gardeners prefer to work with their plants rather than rule them; they seek to escape from a select list of 'cultivated plants', relishing the freedom to pick and choose from those to be found growing wild anywhere in the world. Such freedom comes with a price, however. Many of the plants these gardeners want to grow can seem extremely reluctant to participate in the garden at all. Before this freedom could be used effectively, earlier gardeners had to discover how to germinate seeds of the plants they wanted to grow, then learn enough about the whims and fancies of the plants they produced to enable them to take their place in the garden. We escape from being the servants of seeds only by aspiring to be their masters or mistresses. Gardening may present a gentle face, but it conceals absolute control – often exercised with such subtlety that innocent visitors believe that little planning and less work goes into the making of these glorious displays.

At the age of four or five I badgered my father for a garden of my own, and was 'given' a small corner of the vegetable garden. I spent my pocket money on some packets of seeds with brightly coloured pictures on the front, brought them home and after being shown how to rake the soil into a tilth and draw out shallow drills, tipped the seeds out of the packets along the drills – no doubt pouring out every last seed in my enthusiasm. I obediently raked the soil back over the drills as I was told, and ended by shakily sprinkling water from a can over the whole patch. I was disappointed when I returned after a few hours to find there was not a seedling to be seen, and rushed out before school every morning for the next few days, searching for the first signs of little plants emerging from the soil.

Thousands of others started gardening in just the same way, tempted as I had been by pictures of marigolds, clarkias, candytuft, Californian poppies, pansies and other brightly coloured annuals, and were thrilled, as I was, when flowers just like those in the pictures appeared. Nobody would suggest that run-of-the-mill annuals like these are not amenable to cultivation, nor that they pose any challenge to those who want to master them. Such annuals germinate just as readily and easily as vegetable seeds – for the very good reason that many of them come from the same parts of the world or from places with similar climates, and are the natural possessors of germination strategies which make minimum demands on the gardener's skills.

The Mediterranean Basin, the cradle of so many vegetables, is also the birthplace of numerous, long established garden annuals, among them marigolds, cornflower, larkspur, love-in-the-mist, mignonette, snapdragons, annual chrysanthemums, Moroccan toadflax, stocks, sweet sultan and candytuft. But the Mediterranean Basin is only one of five Mediterranean regions scattered around the world that share a similar climate. In any of them, a visit at the right time of year can leave us marvelling at the exuberance and beauty of the wildflowers.

For thirty days a year Springbok in Namaqualand, some three hundred miles north of Cape Town, is a place of beauty cloaked in flowers; for the other eleven months it is a dusty, heat-ridden, fly-blown, copper-mining town. In the right year, when the rains fall in sufficient quantity at the right time, South Africa, from the Cape of Good Hope to the border with Namibia, becomes a garden filled with the most wondrously colourful variety of flowers (27). Daisies of many kinds – African, kingfisher, Livingstone, Namaqualand and beetle, along with gazanias and ursinias – combine with sun-loving sporries, pietsnots, nemesias and diascias among mesembryanthemums and other plants with succulent leaves. There is also a vast array of bulbs. This magnificent display appears in defiance of a climate that seems sometimes dedicated to the destruction of such lovely things.

California has given us many more wildflowers, and in about 1870 John Muir, perhaps the most inspirational pioneer of the conservation movement in the USA, described the appearance of the San Joachim Valley – now exclusively surrendered to orchards, vegetable and fruit farms – between March and April as 'one smooth, continuous bed of honey-bloom, so marvellously rich that, in walking from one end of it to the other, a distance of more than four hundred miles, your foot would press about a hundred flowers at every step'. Californian annuals may be generally less familiar to gardeners than the long-established natives of the Mediterranean, but their names are no less evocative: tidy tips, goldfields, baby blue eyes, blazing star, five spot, farewell to spring, gilias, Californian poppies and poached egg flowers. Annuals adapted to a Mediterranean environment grow in Western Australia too, among them Swan River daisies (28) and numerous everlasting daisies. And in South America a small area around Santiago with a similar climate has contributed salpiglossis, nasturtiums and schizanthus, the poor man's orchid.

The highlands of Mexico, also associated with so many of the vegetables introduced from the New World, are another prolific source of annuals, although these are not quite so hardy as those from Mediterranean climates. Ageratums come from this part of the world as do cosmos, Mexican sunflowers, French and African marigolds and zinnias.

The fact that easily grown flowering annuals and crop plants and vegetables all come from similar places is only to be expected. The names of corncockle, corn marigold and cornflower tell us they were among the weeds that hitched lifts in Neolithic farmers' seed corn in their long migrations across Europe from Asia Minor. Others, such as sunflowers, ageratum and cosmos from Mexico, are plants which have tagged along with maize. They have been plucked as weeds of the traditional triumvirate of maize, beans and squashes, to be installed in honourable positions in our gardens.

These easily grown annuals are the starter packs for aspiring gardeners, but those of us with any ambitions to speak of soon move on to more challenging things. Perhaps the idea of planting bluebells beneath a group of trees sounds appealing. A neighbour has some plants in her garden and we beg enough seeds to produce several hundred bulbs; a veritable bluebell wood dances in our imagination. Bluebells seed themselves around our woods, so why should they be difficult to grow in gardens?

If we ignore all we have learned to do when growing plants from seeds in packets, listening instead to what the bluebells tell us, they are not. But if we keep bluebell seeds in packets until we are ready to sow them, perhaps the following spring when we sow our lettuces and marigolds, we will have eaten the last lettuce and cleared the remains of the marigolds away long before we see a sign of a bluebell seedling – and will be very fortunate to see one at all. Bluebell seeds, like those of the early forget-me-not and many other winter annuals, mature in mid-summer, lie around till autumn, and germinate as autumn gives way to winter (p. 99). While lying on the ground they take up water and become metabolically active, and during the rest of the summer make preparations to produce seedlings as the earth cools in late autumn. Dry seeds kept in packets, of course, do not respond like those on the soil. Even if we sow the following spring, when the ground is cold, they will produce no seedlings because they have made no preparations to do so. At best, a few may connect with high temperatures during the summer, and a seedling or two may appear the following spring, a year after they were sown; but by then we have probably forgotten about them and used the ground for something else. If we have discovered nothing else, we have been taught why the germination strategies of many plants make them impossible to cultivate in the same way as fields of wheat or rows of carrots.

Success with bluebell seeds depends on sowing them without delay after gathering them. Then, if we poke about in the compost in early winter, we will find little white radicles emerging from the seeds, and can look forward to the appearance of numerous green, slender, cylindrical seedlings by early spring. These seedlings should be transplanted and kept well watered until they die down naturally during the summer. By then they will have formed small bulbs, too immature to produce flowers the following year. But a year later the bulbs will be ready to plant out to make the little bluebell wood that we envisaged. What works for bluebells is equally effective for numerous other bulbs which grow naturally around the Mediterranean, in the Middle East and the western Cape in South Africa.

The seeds of bluebells and other bulbs are not difficult to germinate – just different from those of run-of-the-mill vegetables and annual flowers. So, to a greater or lesser extent, are a great many of the plants that interest gardeners – perennials, grasses, shrubs, bulbs, alpines or trees. Seeds of some come up with no more fuss or delay than annuals or vegetables. Others seldom, if ever, germinate immediately after being sown. They evolved to lie in the soil for a long time preparing to germinate, and will then produce seedlings when the conditions they need have been met and the time is right.

By far the best known treatment for these recalcitrant seeds is a period in the refrigerator – associated most appropriately with plants from alpine regions. Species growing in places where winters are severe are very likely to produce seeds that will not germinate at all, or produce only small numbers of seedlings, before the onset of winter. These respond to low temperatures during the winter and produce seedlings as temperatures start to rise in spring. The temperature increase needed to spur them into growth may be very small indeed, especially in situations where snow lies late and growing seasons are short. Flowers of soldanellas and crocuses in the European Alps, and glacier lilies, shooting stars and Pasque flowers in the Cascade Mountains of Oregon, for example, are familiar heralds of spring, thrusting through the melting snow. Where we see them, we can be sure that their seedlings are not far behind.

The deep-rooted tradition that frost is an essential part of the chilling process is slightly inaccurate. Seeds of most species respond best to temperatures at and just above freezing point, and do so only when fully saturated with water. Putting dry seeds in packets into a refrigerator may keep them alive for longer than leaving them lying around in the potting shed, but it will do nothing to satisfy their need to be chilled before they germinate. Gardeners who grow primulas, gentians, meconopsis, dianthus, aquilegias and globe flowers will find a fortnight or so of frosty weather is usually sufficient to persuade some seeds to germinate, and most will do so after a month. Shrubs and trees may be equally compliant, but for most cotoneasters, philadelphus, acers and birches two or three months are more likely to be effective than a few weeks.

Warm soils in summer prepare bluebell seeds for germination when temperatures fall; cold soils in winter get primulas ready to germinate as temperatures rise. Seeds of those popular garden flowers Christmas and Lenten roses need both warm and cold soils. They are natives of mountainous slopes around the Adriatic Sea, the eastern Mediterranean and Asia Minor – all areas where winters can be cold, and summers, although predominantly hot and dry, also experience periodic spells of heavy rain. Seeds drop from the capsules before mid-summer on to ground that will certainly be warm. It may be wet or dry, depending on whether or not it has rained recently, but with a high probability if it is wet that before long it will be dry, and indeed vice versa. These are not propitious conditions for the survival of small seedlings, and hellebores ensure that none emerge to face them by producing seeds in which the embryos are only partially developed. The embryos cannot complete their development until they have imbibed sufficient water to rehydrate their tissues, which they do during the latter part of the summer. Even when mature, hellebores are in no hurry to face the world (29). Winter lies ahead, and rather than emerge and suffer whatever it brings they lie snug within their seeds, responding to the low temperatures by preparing to germinate when the worst is past. At some point during the winter, usually after Christmas, the seedlings of first one kind of hellebore, then another will emerge, until, well before spring is truly sprung, all have produced vigorous little seedlings.

The vast numbers of seedlings needed to fill a field make it impossible for farmers to cope with seeds that need special treatments before they germinate. However, foresters, who work on a vastly greater scale, though with fewer numbers of plants per hectare, are forced to confront the problem. A high proportion of the plants they grow are propagated from seed. In order to take account of sometimes quite complex, and not always predictable, requirements for conditioning treatments of one kind or another, they have devised simple, straightforward procedures. These meet the needs of most of the seeds they grow, without foresters having to enquire deeply into the exact conditions that persuade them to germinate.

As soon as possible after gathering tree seeds, they are mixed with moist sand or grit and packed into bags. These are put into a pit out of doors, where they are protected from extremes of weather but exposed to natural seasonal changes in temperature. Seeds that have been gathered early are exposed to relatively warm conditions during late summer and autumn, then, as temperatures fall through autumn and drop to low levels in winter, to periods of chilling. In spring, as temperatures start to rise, the contents of bags in which seeds have started to germinate are spread thinly over a seed bed. They are then covered lightly, if need be with more sand and grit, to protect the seedlings as they emerge. Nurserymen grow trees and shrubs, but gardeners can do the same on a small scale by mixing seeds with moist sand, grit or vermiculite, an inert, moisture-retentive mineral, in polythene bags before packing them into a plastic box with a lid in a shaded corner, a garden shed or under the bench in an unheated glasshouse. It is unwise to assume that nothing will happen until spring. Seedlings of many perennials and some shrubs can appear at almost any time, so the plastic bags need to be checked at least once a month for the ivory-white radicles of emerging seedlings.

The principal features of Mediterranean regions are the predictability of the climate, and, sooner or later, of fire. During long hot, dry summers vegetation becomes tinder-dry, and at the mercy of ferocious conflagrations which sweep across the land, leaving desolation in their wake. The aftermath of a fire looks like a disaster zone, but is in fact the prelude to rejuvenation, as the old, tired vegetation destroyed by the blaze is replaced by a mantle of greenery and colour (30). Fresh shoots spring from woody tubers at the base of shrubs, or from buds buried in insulating corky bark. Bulbs produce flowers, seeds which may have lain suppressed for years germinate in their thousands, and seedlings fill spaces cleared for them by the fire.

Gardeners, not surprisingly, concluded that the heat of the fire triggered seed germination. These seeds may have lain in the soil for years, disregarding opportunities to germinate. Bitter experience had shown they could be equally disinclined to produce seedlings when sown in gardens. So, practising what they observed, gardeners sowed seeds in containers, piled a loose topping of straw over them and put a match to it, producing miniature bush fires. The ploy worked sufficiently well to become something of a potting shed parlour trick, but inquisitive gardeners in California and South Africa in the 1950s discovered that the fire's heat is only part of the story. The cones, capsules, follicles and other woody fruits of some pines and gum trees, as well as Australian woody pears, banksias, bottle brushes and hakeas, do indeed only open to release their seeds after they have received a toasting. However, it is not the heat to which the seeds themselves are responding when they produce seedlings, but the complex mixture of volatile organic chemicals in the smoke.

Such observations put a dampener on potting shed pyrotechnics. Californian gardeners took to the blow torch instead, torching the twiggy shoots of shrubs and then grinding the charred remains into powder, known with simple logic as charate. They sprinkled charate over the surface of the potting mix in which they had sown seeds of different kinds of chaparral plants with gratifying results, attributed to the action of volatile chemicals washed on to the seeds.

Hannes Lange, while working at the National Botanical Institute at Kirstenbosch in South Africa, tried a different approach on species of the heath-like vegetation typical of the region known as *fynbos*. He arranged for smoke to blow over the pots in which he had sown seeds, with some remarkable results. Among the most interesting were a group of sedge-like plants known as restios, which had previously proved particularly intractable and which responded by germinating with unhoped-for abundance. The smoke-blowing idea was taken up by his colleague Neville Brown, who devised means of persuading gardeners to buy smoke by impregnating small pads of absorbent paper with it. Meanwhile Terry Hatch, a practical-minded nurseryman and the owner of Joy Nurseries in Pukekohe, New Zealand, used a home-made fish-smoker to persuade various bulbous plants and their seeds that they had experienced a bush fire.

Fire is also believed to improve the germination of some seeds, including lupins, gorse and brooms, by cracking open their seed-coats – a prospect that, bearing in mind the effect of a little roasting on popcorn, sounds highly plausible. However, fire is not the only means by which the hard shells of seeds can be breached. The hard shells of legume seeds prevent them from producing seedlings, which frustrated gardeners until they came up with a simple and direct solution to the problem. They use a knife to chip away a small part of the shell, which breaks the seal and allows water and oxygen to reach the embryos. It is clear, therefore, that germination strategies can involve numerous different factors that may often interact in complex ways.

When Roger Smith and I returned from a seed collecting expedition in southeastern Europe to the seed bank at Wakehurst Place in 1974, we brought numerous collections of vetches, clovers, medicks and other members of the pea family. Flushed with success, we did not expect to have trouble germinating their seeds. They were duly sown, after chipping their seeds coats in the approved manner, and in a short time healthy seedlings appeared. A week or two later these stopped growing and turned yellow, before collapsing and dying. We seemed to be afflicted by a frustrating and puzzling problem until, more belatedly than should have been the case, the obvious answer presented itself. The roots of plants in the pea family are unable to take up nitrogen for themselves, and depend on alliances with bacteria living in nodules in their roots to assimilate nitrogen from the air.

We had forgotten that seedlings need to do more than emerge at a favourable time and avoid becoming lunch for one or another of innumerable predators. Many also have a third hurdle to clear on the way to a flourishing future. They must find bacteria or fungi with which to form mutually beneficial alliances, known botanically as symbiotic associations. Under natural conditions, the roots of our seedlings would have quickly been invaded by bacteria present in the soil. This invasion would be controlled within the roots by limiting the bacteria to cells located in nodules within which they lived – processing nitrogen for their own use and paying rent by releasing some to their host. Our seeds, sown in a sterilized potting mix, had been unable to find the bacteria they needed. They died because once the little supply of protein in their seeds was exhausted, they had no more upon which to live. We solved the problem by mixing a little meadow soil with the potting mix before sowing the next batch of seeds.

Old-fashioned gardeners, undeterred by notions of hygiene now taught so assiduously on horticultural courses, were not picky about cleaning up their potting benches. Some believed that a little old potting mix added to the new actually improved prospects when sowing seeds or pricking out seedlings. Such slovenly behaviour in the potting shed would earn few marks on a diploma course today, but, as we discover more about the array of mutually beneficial alliances between plants, fungi and bacteria, the idea begins to make more sense. Gardening routines prepared in obedience to the goddess of hygiene's demands are inimical to symbiotic alliances. Strict attention to cleanliness denies seedlings opportunities to interact with soil micro-organisms. Fungicides used routinely or unnecessarily prevent symbiotic relationships by destroying friendly as well as hostile fungi and bacteria.

Nodules on the roots of plants of the pea family are so conspicuous they have long been well known. But alliances between plants and bacteria range far further than the species of *Rhizobium* associated with peas and vetches. Thread-like bacteria in the much less familiar genus *Frankia* are associated with a wide variety of plants, including alders, ceanothus and coriarias. Even less familiar are ancient associations between cycads and thread-like bacteria known as blue-green algae which, unlike *Rhizobium*, can fix nitrogen in the air.

Alliances between plants and bacteria or fungi were once thought more likely to be the exception than the rule. Now we know that all but a minority of plants form symbiotic associations with fungi, and almost all have established relations with bacteria of some kind. We now know that plant chloroplasts, the part of the cell where photosynthesis takes place, have evolved from symbiotic associations with cyanobacteria.

Numerous different fungi, all members of a particular group called the Glomeromycota, live perpetual, clandestine lives inside the cells of roots. The mutually beneficial association between the fungus and host plant is an endomycorrhizal one (literally 'fungus inside a root'). Here the fungi develop as congested bundles of threads known as arbusculae – or little bushes – because of their appearance. Under natural conditions these endomycor-rhizal associations are believed to exist in at least 80 per cent of plants worldwide. Gardeners seldom become aware of their existence, though their presence can often be detected in long established perennials and shrubs by the characteristic claw-like appearance developed by the roots they inhabit. In orchids endomycorrhizal associations are restricted to a single fungal genus, *Rhizoctonia*, which enters their germinating seeds in the earliest stages of development and supports their early growth and later metamorphosis into fully fledged plants.

We see mushrooms, fly agarics, chanterelles, truffles, ceps, boletes and other toadstools. However, we do not see the fungal hyphae, the threads that make up the fungal body. The hyphae spread out like branches through the soil, shrouding roots and invading spaces between cells within the plants, though not actually occupying the cells themselves. Known technically as ectomycorrhizal fungi, toadstools live in close-knit alliances with numerous trees, shrubs, grasses and perennial plants.

A third group, the ericoid mycorrhizal fungi, were once believed to be confined to the roots of heathers and related plants, including Australian species of *Epacris*. Their discovery in liverworts suggests ericoid mycorrhizal fungi are much more catholic in their choice of friends than once thought, and may even form complex triple alliances involving flowering plants, fungi and liverworts.

As we discover more about the alliances formed between plants, fungi and bacteria, their importance becomes steadily more apparent, and the world in which the roots of plants exist more and more fascinatingly complex. The parts that fungi and bacteria play in fixing atmospheric nitrogen and making the less soluble sources of phosphorous in the soil available have long been recognized. Now an extraordinary story is unfolding to reveal the subtlety and diversity of the alliances and conflicts that go on beneath the soil surface, involving numerous microscopic soil organisms.

Roots are constantly vulnerable to pathogenic bacteria and fungi, which can overwhelm and destroy them. The presence of a friendly organism not only shields them from these attacks, but also protects them from toxic substances which can accumulate in the confined spaces within the soil. Alliances with ectomycorrhizal fungal hyphae greatly increase the volume of soil from which plants can obtain water and nutrients, and the combination of better protection and increased foraging efficiency greatly improves a plant's chances of competing effectively with its neighbours. This is of lesser value in the protected conditions of a garden, but under natural conditions it spells the difference between survival or rapid elimination.

Orchids, as we have seen, prefer exclusive relationships of their own. Their seed capsules may be packed with hundreds of thousands of seeds, each consisting of a minuscule embryo composed of a bundle of undifferentiated cells enclosed within a filamentous net and so lacking in substance that they waft away on a breath of air. Orchids launch their seeds into the world in a state of destitution, so deprived of resources that unless a *Rhizoctonia* species is at hand to sustain them, they have not the slightest prospect of a future.

That did not prevent gardeners finding ways to grow orchids from seed, an enterprising discovery which received its just reward when it transpired that not only were orchids careless of their offspring's future, but also extremely promiscuous. Plants in general hybridize only with closely related species; even then sterility barriers frequently intervene to prevent the formation of hybrids, or any hybrids produced are sterile. Orchids, on the other hand, hybridize with an abandon that seems indifferent to close relationships, and hybrids between different genera of these flamboyant flowers, even between three or four different genera, are not unusual.

This promiscuity caused considerable heartache at Kew during the 1960s, when the living collections were being radically reorganized to weed out plants of doubtful or hybrid origin. Exotic orchid hybrids were liberally represented in the glasshouses, occupying space needed for plants of known wild origin. Over the years numerous widows had donated their ex-husbands' cherished, and often very valuable, collections of hybrid laeliocattleyas, odontiodas, cymbidiums or fashionable dendrobiums, miltonias or paphiopedilums to Kew. So to Kew they came, where these flamboyant, showy and often gorgeously beautiful exotics now played the role of cuckoos as they displaced the less floridly theatrical orchid species that had become the favoured darlings of the new thinking.

The obvious way to obtain the species needed to renew the collections was to collect them from the humid, tropical jungles or cloud forests where they grew naturally. But commercial collectors, as well as botanists, had done this too often and with too much enthusiasm in the past for it to be a viable option. Stripping the trees, then packing orchids off indiscriminately and unceremoniously in the hope of financial gain or scientific kudos was now regarded as something respectable botanists preferred not to be observed doing. In only a few years the importation of wild orchids became forbidden by an international agreement embodied in the Convention of International Trade in Endangered Species of Wild Fauna and Flora – referred to more conveniently as CITES.

Since senior botanists at Kew were among those most active in pressing governments to support the Convention – and had indeed taken a leading part in the drafting of its clauses – opportunities to collect plants from the wild were, generally speaking, better not pursued. This would have made more sense if the forests in which the orchids grew were not being destroyed wholesale to obtain timber, or to clear land for ranching or growing crops. However, even rescuing orchids doomed to die as the trees were felled carried too many overtones of past rapacity, and was subject to too many ambiguities to be pursued. Seed collecting was recognized as the acceptable way to obtain orchids and, since a single orchid plant produces millions, or tens of millions of seeds in its lifetime, one which could have no significant impact on the survival of plants in the wild.

Hitherto few attempts had been made to grow orchids from seed at Kew, apart from haphazard procedures during which seeds were sprinkled on top of the potting compost in the pots where their parents were growing. Sometimes this worked, but quite often any seedlings which appeared were those of one or two vigorous species which sowed themselves around unbidden, and were regarded as weeds.

Nevertheless, orchids had been more or less regularly grown from seed in this way ever since David Moore, director of the Glasnevin Botanical Garden in Dublin, wrote an article in *The Gardeners' Chronicle* in 1849. Moore described his success in growing seedlings of epidendrums, phaius and cattleyas by sprinkling their seeds over the surface of the compost in pots already occupied by their parents, or over pots filled with compost previously occupied by orchids. Most encouragingly he was able to report that seedlings of epidendrum and phaius flowered three years after being sown.

Orchid growers quickly latched on to the possibilities, and throughout the latter half of the nineteenth century, and well into the twentieth century, numerous hybrids were produced by following Moore's methods. The collections of orchid fanciers filled with exotic creations as growers devised their own jealously guarded elaborations of the method – some practical, some weird, some wonderful, and some even useful. All of them were based on finding effective ways to nurture unions between the orchid seeds and their fungal associates. During the first years of the twentieth century two scientists, Noel Bernard in France and Hans Burgeff in Germany, put things on a more scientific basis when they discovered how to germinate orchid seeds in test tubes on sterile nutrient media inoculated with the *Rhizoctonia* fungus. Generally the orchid growers preferred the practical atmosphere of the glasshouse to the laboratory and went on producing hybrids as enthusiastically as ever in their own way – with one notable exception. Joseph Chamberlain, a nurseryman with no scientific training, adopted Bernard's methods, and produced thousands of seedlings of odontoglossum hybrids on his nursery near Haywards Heath in Sussex.

A couple of decades later, an ingenious and versatile botanist called Lewis Knudson, based at Cornell University in Ithaca in New York State, had a notion he might be able to make the fungal partner redundant by supplying orchid seedlings with whatever nutrients it contributed. Eventually, and despite loudly voiced doubts from many fellow scientists, he managed to germinate and grow orchid seedlings in glass flasks by sowing them on a jelly containing sugar and a full range of mineral nutrients. When they grew large enough he removed them from their flasks and set them out in pots of orchid compost in which they eventually developed into flowering plants – just like those which had enjoyed the benefits of natural symbiotic associations.

Lewis Knudson showed the way, and before long the laboratory took over from the potting shed as the place where orchid seedlings were grown. But Knudson was a pioneer of even more significant discoveries. He had shown it was practicable to grow plants under artificial conditions in laboratories, and this led to an increasing interest in the possibilities of propagating plants of all kinds by similar methods. In the 1950s the physiologist Frederick Steward, also working at Cornell University, removed a single cell from the root of a carrot, and grew from it a normal carrot plant, complete with flowers. In doing so he demonstrated a phenomenon known as totipotency, a word coined to express the fact that the nucleus of every cell contains all the information needed to produce an entire plant.

Totipotency makes the gardener's practice of propagating plants from cuttings possible. Thus, cuttings made from sections of root are capable of producing shoots and flowers, and small shoots, detached as cuttings, develop roots. Although in normal circumstances cells in roots only form organs and tissues appropriate to roots, and cells in shoots develop into the organs and tissues of shoots, any cell is capable of forming organs of any kind when the need arises. The concept of totipotency has been extended and refined until today a very high proportion of the shrubs, perennials, bulbs and roses we buy in garden centres are propagated from cultures of cells in commercial laboratories. These are developed under very similar conditions to those first used to grow orchid seedlings during the 1920s.

When the decision was made to grow orchids from seed to replace the hybrids, now out of favour, no one at Kew had any experience of laboratory-based methods. A joint project was thus set up in the late 1960s between the newly established Physiology Department and staff of the Tropical Section of the Gardens Department.

Initially work was done in the Jodrell Laboratory, where I was most fortunate to be joined by Udai Pradhan, owner of a notable orchid nursery near Kalimpong in Northern India. Udai's keen interest in orchids was equalled by his remarkable gift for growing them. We first needed to find out if we could use standard nutrient solutions, such as those developed by Lewis Knudson, to grow any species of orchid, or whether different species had different requirements. My previous experience of using mineral nutrient solutions, while working at Long Ashton Research Station near Bristol, combined with Udai's enthusiasm for the project, quickly led to the development of a new range, known as the Jodrell Orchid Nutrient Solutions. These had the unique advantage of enabling us to vary the concentrations of different nutrients individually in order to study the needs of different species.

24 Maple fruits are seen spinning like helicopter blades as they fall from the tree. Since the speed with which the fruit falls is lowered, there are ample opportunities for wind to catch and disperse the fruit.

25 The coconut, so characteristic of tropical beaches, disperses its fruits by ocean currents. A coconut fruit, with the seed having germinated inside, is shown tossed up on a tropical beach.

26 Pumpkins and squashes are grown in gardens in a remarkable array of colour, sizes and shapes. These fruits are the products of centuries of selection by humans.

27 The Cape region of South Africa is one of the world's most diverse environments. During the flowering season the landscape is a mosaic of yellow and orange daisies, and purple heathers.

28 Swan River daisies are annuals adapted to the Mediterranean environments of Western Australia. The daisy 'flower' is in fact many tiny flowers packed together in a head. Each tiny flower produces a single seed.

29 The seeds of hellebores are contained inside fruits surrounded by sepals that persist long after the petals and other parts of the flower have disappeared. Hellebores are plants of regions with cold winters and hot summers, so seeds fall onto warm ground but must survive the winter.

30 Fire has passed through this area leaving dead trees. However, on the ground grasses and other herbs have started to germinate. Fire is an essential part of many habitats, which leads to rejuvenation of the landscape.

We were joined in the early 1970s by Dickon Bowling, from the Gardens Department. During the next three or four years we grew seedlings of more than one hundred and thirty different species of tropical epiphytic orchids, most of which had never been grown from seed before. The results conclusively proved that this was a practical and effective way to produce the orchids needed. Soon afterwards a plant propagation unit was set up at Kew to produce plants for the collections in more innovative and technical ways than those traditionally used.

Decades later, long after my own involvement with the project was over, I encountered an aspect of its consequences during a visit to Madagascar. While returning to Antananarivo, Madagascar's capital, after a visit to the Analamazoatra Reserve near Andasibe, we passed a woman standing at the roadside beside half a dozen sections of tree fern stem, each with an orchid plant growing on it. Our driver got out, ran back and returned with four large orchids, for which he had paid the woman a pittance.

Hawking orchids is one way in which the villagers make a little money. It is, unsurprisingly, deplored by conservationists, who argue that the orchids have been taken from trees in the forests and should be left where they grow naturally. Some may have been rescued from trees felled for legitimate purposes, such as charcoal burning, but of course, many have not. Madagascar's impressive orchids have suffered drastic reductions in range and numbers with the destruction of the forests and, again not surprisingly, this is a sensitive area.

A sensitive area indeed, but not only for those concerned about conservation. Orchids are one of the natural resources from which the people of Madagascar make a living. There are many women like the one we had just met. They do not sell enormous numbers of orchids, and their sales are insignificant by comparison with the losses caused by destruction of the forests. Nevertheless, the money they make from these sales can be almost their sole source of income.

Kew has been involved in projects to conserve some of the thousand or so Madagascan orchid species, including growing orchids under laboratory conditions for introduction to Madagascar. However, Kew's project would not help the woman selling orchids by the roadside. It may reduce pressure on wild orchids, but the more successful it is the more it will reduce her already meagre sales – seen as a successful outcome by those concerned only with conserving the country's orchids. One of the more intractable problems of conservation is finding ways to compensate local villagers for the losses they suffer when they are excluded from entry to reserves or denied the benefits of the often life-supporting resources of plants and animals, upon which they previously depended.

Laboratory facilities and equipment are immeasurably beyond the most extravagant hopes of impoverished villagers. However, the vast numbers of orchids produced during the century after David Moore had scattered seed over compost impregnated with the fungi that were beneficial (symbiotic) to the orchid proved that there was another way. I should very much like to see a revival of interest in these methods, so that villagers in Madagascar – and many other countries in the tropics – could be shown how to grow orchids as a cottage industry. This could be done with few resources and virtually no capital. It would enable those least able to help themselves to benefit by working with, rather than against, conservationists, instead of adding to the growing tally of examples of well-meant conservation projects that turn out to be against the interests of the poorest members of the community.

Perhaps the last word on the objectives of plant conservation should go to Etienne Rakotomaria. In a speech welcoming delegates to a conference in Antananarivo organized by the International Union for the Conservation of Nature in 1970 he stated, 'The people in this room know Malagasy Nature is a world heritage. We are not sure others realize it is our heritage.'

CHAPTER SEVEN

The Pursuit of Plenty

No other vehicles were on the road; instead people and animals stretched ahead into the distance. Albanian, Macedonian and Bulgar villagers from valleys tucked away in the surrounding mountains were returning home from market in Debar, a town in what is now Macedonia. Five, gleaming black, hairy water buffaloes ambled ahead of us, trailing obediently behind a small girl whose head barely reached halfway up their flanks. A man escorted a large mottled sow to a new home, prodding errant piglets back onto the straight and narrow with a long bamboo cane. Groups of peasants perched on sacks of provisions, loaded on horse-drawn wooden carts, enjoying the cool of the evening after a long hot day. Men with bags of horn-shaped yellow capsicums slung from their saddles rode on ponies, while women, trudging beside their mounted husbands, shepherded groups of children through the haze of dust above the road.

In the fields weary reapers piled the last sheaves of the day onto cumbrous farm carts with heavy timber wheels and sides of rough-hewn planks. Oxen, heaving and shoving one against the other, ponderously urged carts into motion across the uneven ground and dragged them out onto the road. Men still scythed in a few of the fields, attended by women who gathered the long straw in their arms and deftly bound them into sheaves. One of the tumbrels turned into a small enclosure at the outskirts of a village, in which half-made ricks surrounded a circular stone platform. Later the ricks would be taken down and the straw, with the grain still in the ears, spread over the platforms. Donkeys or oxen, drawing heavy wooden boards behind them, would walk round and round to dislodge the grain, ready for the men to complete the process with their flails. After the straw had been bundled up, the grain would be retrieved by throwing it, together with the chaff, into the air. Wind would then carry away the dross, while the fully developed kernels fell back to the threshing floor.

My calendar said it was 1 August 1970; my eyes told me it was a thousand or more years earlier. In this corner of Macedonia, pinned against the mountains of Albania, the familiar actions of market day and harvest time had followed the same pattern since time immemorial (31). Muscle power – human's or beast's – tilled the soil and gathered crops grown from home-saved seed. That was where seed had always come from. It was simple, straightforward and cost nothing. From time to time, of course, neighbours would exchange seeds. Occasionally replacements would be bought in the local market, and over the years each district, each village, each farm even acquired its own particular strains of cereals and other crops. They were selected, inadvertently, but precisely, to match the climatic peculiarities of the place and the individual needs of people with different cultures and tastes.

The peasants used their crops in a variety of ways. Generally speaking, wheat, barley, rye or oats were grown in separate fields, but the farmers had learned the dangers of being too dependent on any of them. Quite often there was a casual mixture of three or even four different cereals, sometimes with peas, lentils and even buckwheat growing among them for good measure. The wheat and barley reached to the men's waists, and reapers in the fields of rye were almost hidden from view amongst the standing straw.

People had learned to live with both the good years and the bad. They had known years of plenty and years of shortage – and from time to time abysmal years when drought, pests or diseases decimated the crops and famine hovered in the background. Yet thrifty, low yielding kinds of barley or rye were able to survive droughts which reduced their more luxuriant companions to thin shadows. Scrawny specimens of primitive wild wheat or oats would escape infestations of insects or mould infections that thrived on their fatter cousins. The landraces were a miscellany of mongrels, composed of infinitely varied combinations of genes. These glorious mixtures yielded less than pedigree strains in favourable conditions, but in adversity, when rains failed, bugs bit or fungi flourished, somewhere among them would be plants with just the tolerance to this or resistance to that needed to produce a small crop. In parts of the world where people's lives depended on subsistence in bad years, that was much more important than the extra yield of pedigree strains under favourable conditions. The resilience of the ancient landraces provided a natural insurance against adversity. It was needed by farmers forced to accept situations and conditions as they found them, because they had no power to change things.

Five days later we were in Greece, and agriculture had leaped forward two thousand years or more. Fields of wheat and barley stretched across the Plain of Thessaly from the sea to the lower slopes of the mountains in disciplined array. Each densely packed field was filled with uniformly sturdy plants topped by impressively stout ears, with never a plant of one kind intruding into a field intended for another.

The two scenes might have been dioramas, arranged by a museum curator to illustrate the impact of events described in the first two chapters of this book. The peasants in Macedonia still followed the old ways, raising their own crops for their own purposes according to traditions handed down from generation to generation: a situation not so very different from the conditions under which plants had originally been domesticated. In Greece the ancient inheritance of landraces and crop diversity had been swept away in the course of less than twenty years by the wholesale introduction of improved pedigree strains of wheat and barley. In this chapter we will look at the people, places and events that transformed farming, almost throughout the world, from largely disorganized and diverse forms of crop production to highly coordinated agribusinesses, based on a select number of high yielding strains of all the major crops.

The farmers in the villages around Debar paid no attention to genetic diversity, a modern term which would have meant nothing to them at all. They had no need to, because after we had learned how to grow our own crops of barley, wheat, rice, maize or whatever, Nature supported us. In the course of discovering how to grow crops, our ancestors had also been provided with an almost foolproof recipe for survival, if not for abundance. By the middle of the nineteenth century a better understanding of the possibilities that seeds provided for crop improvement, and a keener awareness of the slender basis on which our survival depends, tempted us to desert the tried and tested methods in the pursuit of plenty.

Gregor Johann Mendel may not have been inclined to work on his father's peasant holding in Moravia (p. 59), but he would nonetheless have been at home in the fields of Macedonia. Both the appearance of the fields and the activities of the harvesters would have been much as he remembered them in the mid-nineteenth century. But Mendel would hardly have recognized the uniform fields in Greece, where plants, barely up to his knees, produced large heads of plump grains – and would have been utterly astonished to learn they had any connection with his experiments on peas all those years ago in the Monastery of St Thomas in Brno.

By the beginning of the twentieth century, barely forty years after publication of Charles Darwin's *On the Origin of Species* and the details of Mendel's experiments on peas, there had been a sea change in attitudes to the natural sciences. Theology and philosophizing no longer provided the basis for interpreting the behaviour of plants and animals, partly because tenured posts in the universities had ceased to be the preserve of clerics. This was due not least to the activities of hundreds of amateur naturalists, clergymen, country gentlemen and others whose contributions underpin our knowledge of natural history, geology and archaeology, and partly due to widespread secularization of research into these topics, complemented by the willingness of private individuals to support, and fund, research.

In England alone, for example, John Bennet Lawes set up an agricultural research centre in 1843 on his estate at Rothamstead in Harpenden, some twenty-five miles north of London. In 1876 Thomas Jodrell Phillips-Jodrell, a friend of Joseph Hooker, director of the Royal Botanic Gardens at Kew, provided the funds to build and equip a small laboratory devoted to the study of experimental aspects of botany there. Some thirty years later, the South Eastern Agricultural College started life at Wye, in Kent, on the site of an ancient educational establishment dating from 1447. During the 1890s Robert Neville Grenville started experiments on cider making at his home near Glastonbury, and was subsequently a prime mover in setting up the National Fruit and Cider Institute near Bristol in 1903, in an old cart shed and chicken house; in due course it became Long Ashton Research Station. A few years later a bequest from John Innes, a City of London merchant, funded the institute at Merton in South London that bore his name, and in 1913 nurserymen and orchard owners in Kent set up the East Malling Research Station to support the local fruit growing industry.

All these establishments would later become a charge on the taxpayer when government assumed responsibility for them, but in their beginnings they were provided for by private individuals, supplemented in a few cases by modest donations from neighbouring county councils. Similar research institutions had been set up in many parts of Europe, in the USA and across the British empire. The study of plants was no longer an academic pursuit of university dons, but part of the rapidly developing sciences of agriculture and horticulture in which research had become a tool, aimed at finding solutions to practical problems.

Before the end of the nineteenth century, the days of the landraces were over in the more developed parts of the world. Artificial fertilizers were gradually becoming accepted. Many pests and diseases could be controlled with chemical sprays. Sophisticated systems of crop rotation had been devised to preserve and enhance the fertility of the fields. Pedigree strains of cereals, arable crops and vegetables produced by plant breeders were yielding greater and better quality crops than had been possible only a few decades previously. Plant growers no longer had to accept conditions as they found them: they were acquiring the power to make changes, to improve situations and to remedy problems. Power led to greater expectations, and amongst their expectations were higher yields and better quality products from all the crops they grew.

The old landraces were condemned as 'peasant agriculture' with no place in the modern world – and with some reason. It had taken nearly a thousand years for wheat yields to rise from half to two tonnes per hectare in southern England. Within forty years the introduction of improved, pedigree varieties and better cultural methods would raise expectations from two to six tonnes. The capacity of the landraces to produce a meagre sufficiency even under dire conditions lost its appeal in an age when science promised to eradicate such dire conditions. It was obviously more profitable to grow pedigree, uniform strains with an exclusive representation of the genes that gave high yields. So what if they did provide less security in adverse conditions? Adverse conditions were something to be remedied, not endured. Yet by switching from landraces to pedigree strains farmers and gardeners had cut the apron strings provided by a supportive Nature for their security. The future now lay in their hands. As the nineteenth century ended and the twentieth began, those hands would be further strengthened by the application of scientific principles to what had formerly been the empirical practices of plant breeders.

In 1899 Gregor Mendel was rediscovered. Thirty-three years after the non-event of the publication of his experiments on peas, and fifteen years after his death should have laid them to rest in peace as part of the vast, unremembered corpus of discarded scientific ideas, fate intervened. Within ten years Mendel had been elevated to the pantheon of founding fathers of a science – the science of genetics. This was not because his discoveries were any longer revelations, nor because they illuminated the way ahead for others striving to understand the mysterious laws of inheritance. It was not even because of the intrinsic value of the truths that Mendel had revealed, significant though these were. Mendel achieved his belated celebrity through the foibles, jealousies, ambitions and craving for recognition of four academics, each of whom had picked up the long-cold scent leading to the secrets of the laws of inheritance.

Hugo de Vries, director of the Botanical Institute in Amsterdam, conducted experiments on maize and a dozen or so other plants that had come up with results closely similar to those obtained by Mendel. But did de Vries already know about Mendel's work, before he started his experiments? Had he yielded to temptation and not declared the source of his knowledge when he published his results in 1899, in the hope of obtaining credit for himself? Or had de Vries come across Mendel's paper when his own work, the fruit of his own inspiration, was almost complete, and yielded to the very human feeling that he was entitled to the credit for it? Either way he was forced, belatedly, to acknowledge Mendel's precedence in a postscript to the publication in which he announced his results.

Carl Correns was lecturer in Botany at Tübingen University and a former student of Karl von Nägeli, professor of botany at Munich University and a correspondent of Mendel's for a number of years. Correns's experiments on maize and peas had corroborated those found by Mendel, with some significant exceptions. In 1899, as he was preparing a paper on the subject, he discovered he had been pipped at the post by de Vries. He promptly fast-forwarded the publication of his paper, taking care in doing so to draw attention to the common debt that both he and de Vries owed to Mendel. Was his action in order to deprive de Vries of victory, preferring to share silver rather than see his rival make off with the gold?

The third man was Erich von Tschermak-Seysenegg, at the time a post-graduate student based first in Ghent, then the University of Vienna; he was later to become an eminent plant breeder. Von Tschermak also published the results of experimental studies on peas during 1899 and made due reference to Mendel's publication of 1867, although nothing in his text suggests that he had grasped the significance of Mendel's conclusions. Like the others, he added a postscript in which he staked his claim to be one of Mendel's discoverers. It is not known whether von Tschermak sought to climb on the bandwagon when he realized there was much to be gained by doing so, hoping to join the list of scientific notables by being able to claim, 'I said that too.'

William Bateson was the fourth man, a Fellow of St John's College, Cambridge. He published nothing about Mendel in 1899, but his meeting with Hugo de Vries at the Royal Horticultural Society's First International Conference on Hybridisation and Plant Breeding in that year alerted him to what was afoot. During the next few years Bateson ruthlessly built on Mendel's legacy to establish the science of genetics. He seems to have been a man whose actions were largely governed by an inability to let any oppor-tunity of scoring over those whom he saw as his rivals slip. His motive may have been a selfless recognition of what the world owed Gregor Mendel; expectations of his own self-aggrandisement as the high priest of the new science may also have played a part. Perhaps the satisfaction lay in reducing the reputation of Hugo de Vries, and all others who were, or were deemed to be, not of Bateson's mind.

The answers to such speculation form a tangled tale too long, complex and open to different interpretations to go into here. Whatever the motives, the outcome was Mendel's elevation to heights to which he could never have dreamed of aspiring – established as the founding father of the science which deals with the ways characters are inherited, and which underlies the entire plant breeding effort of the twentieth century. The enquiring, iconoclastic minds of nineteenth-century botanists had uncovered the hidden possibilities within seeds, revealing opportunities to breed higher yielding varieties. Scientific research during the twentieth century would build on this platform, turning opportunities into realities beyond the dreams of those who first contemplated it.

But science did not monopolize all the pegs. Before we get immersed in technicalities and mammoth enterprises on national and international scales, it is worth remembering that there was, and still is, room for the innocent amateur. The story of George Russell, who did it all on an allotment in the city of York, is an encouraging reminder that anyone brave enough to follow his muse can breed flowers capable of lightening gardeners' spirits throughout the world.

Lupins grow wild in damp meadows along the sides of streams in British Columbia, Washington and Oregon (32). Their flowers are either pink or purple, and most gardeners in the early twentieth century regarded them as attractive rather than inspirational. However, George Russell's infatuation led him to commit the ultimate heresy of filling his allotment with them.

George Russell was a taciturn man. He said little, but he began to wonder whether he could add more colour to his purple or pink lupin flowers. What about the yellow lupins that grow naturally on dry, free-draining sand dunes along the coasts of Oregon and Washington? During 1911 Russell planted some seedlings in his allotment. The bumblebees did the rest. They flew from flower to flower, transferring pollen from purple flowers to yellow and from yellow to pink. Russell collected the seeds and sowed them. When the seedlings grew up, some had flowers that combined the colours of the purple and pink kinds with the yellow. He pulled up those he did not like, and the bumblebees continued to move the pollen from one flower to another. Russell, who knew nothing of plant breeding, simply kept the best and dumped the rest.

Year by year the bumblebees worked and Russell selected, until his allotment was filled with lupin flowers in every combination of purple, pink, yellow, tan, apricot, cream and white – some with uniformly coloured spikes, others with bi-coloured harlequin combinations, unlike anything seen before (33). He obstinately refused to sell a plant, or even a seed, to anyone. For once he was obeying the rules. Allotments were places where fruits and vegetables were grown strictly for home-consumption, enshrined by convention often reinforced by strict regulations posted on a notice board by the gate.

When George Russell was in his eighties his wonderful, multi-coloured lupins attracted the notice of Jimmy Baker, proprietor of Bakers Nurseries near Wolverhampton. A nurseryman with a keen nose for business and an adventurous spirit, he proved to be just the partner George needed. Baker quickly found a way round the interdiction on selling allotment produce by transferring George's lupins to his nursery, then offering him a deal he could not refuse for the right to market them.

The lupins were presented to an appreciative public on Baker's stand at the Royal Horticultural Society's monthly show in June 1937. It was coronation year, and in the previous month King George VI and Queen Elizabeth had ascended the throne. Yet for many in the gardening world, the high point of the year was the debut of George Russell's lupins.

George Russell was quite happy to plant his lupins, pull out those he did not like, but otherwise let Nature take its course, which it did with remarkable effects. Like the old landraces of wheat and barley, genetic diversity was an integral part of the strain of lupins he produced. Sophisticated modern plant breeders do not work like that. Their aim is to produce pedigree varieties in which genetic diversity is reduced to a minimum – ideally each and every plant should possess identical combinations of genes, carefully selected and composed to produce a crop tailored to the exact needs of the grower. Genetic uniformity becomes the ideal. Genetic diversity, unless preserved separately and deliberately, gets lost along the way.

By the end of the nineteenth century, Russia and the USA had set up state-run plant breeding programmes, backed by extensive seed collecting expeditions at home and abroad. These enabled them to gather seeds of crop plants, vegetables, fruits and other useful plants, and make them available to their plant breeders in their attempts to produce new and better varieties.

One of the pioneers was Robert Regel, director of the Bureau of Applied Biology in St Petersburg around the turn of the twentieth century. He was responsible for assembling collections of crops in Russia and further afield, often in cooperation with collectors working for the USDA. One such was a plant breeder employed by the department. Mark Carleton visited Russia during the 1890s in search of varieties of wheat capable of producing satisfactory crops in semi-arid districts of the Great Plains. His search was successful, and several of his collections performed as hoped, but a few years after Carleton's return a serious outbreak of black stem rust decimated crops of wheat across large areas of the States. Quite unexpectedly, some of the Russian collections remained scarcely affected by the disease. This was a tremendous bonus, as black stem rust was practically uncontrollable at the time and repeated epidemics had become a major problem. Now Carleton had produced new varieties which were not only drought tolerant, but rust-resistant as well. There was a snag, as the desirable wheats were all durum varieties – good for pasta production but unsuitable for use as bread wheats. Further crosses and trials were needed to combine the tolerance and resistance of the introductions with the high gluten levels needed to produce good dough. Once that had been achieved, the way was open for the development of the economically crucial resource provided by the wheat belt of the northwestern states.

Robert Regel was succeeded in 1917 by Nikolai Ivanovich Vavilov – the number one candidate for patron saint of plant genetic resources. Russian farming had for generations run on the backs of the *kulaks*, or peasants, and this continued following the Revolution. Kulaks relied largely on seeds collected year-on-year from their own fields, and which their forebears had collected before them. Between them the peasant farmers possessed a seemingly almost infinite, minutely adapted range of wheat, rye, barley, oats, buckwheat, sunflowers and other agricultural crops, selected over the years for the conditions of the places where they were grown.

Vavilov built on the foundations prepared by Robert Regel to construct one of the most comprehensive enterprises ever seen in the world of botanical endeavour. He recognized the genetic diversity held in the landraces as the key factor in breeding the new varieties of cereals and other crops required to meet the hugely different conditions found within the Soviet Union. In order to do this Vavilov embarked on a series of expeditions to collect as many different varieties as possible of cultivated wheat, barley, maize, potato tubers, beans, fodder and forage crops, fruits and vegetables, as well as their weedy associates. His enterprise extended far beyond just making collections. Plants were grown on trial in different places to assess their good points and drawbacks. Samples of seed were stored to make sure they remained available for future use. An extensive plant breeding programme was set up to put the material to practical use.

His expeditions covered the whole of Soviet Union, including the recently acquired countries of central Asia. This would have satisfied the ambitions of most men, but not Vavilov and his colleagues, who then set out to cover the rest of the world. They eventually collected seeds in more than sixty different countries, even establishing an office in New York during the 1920s, through which he enrolled Russian emigrants to make collections for shipment back to the Soviet Union. That did not pass unnoticed by the commissars of the USSR, and aroused deep-rooted suspicions of his reliability among members of the communist hierarchy.

Even while Vavilov and his colleagues were making their collections, their significance would be underlined by events. The Bolshevik revolution of 1917 transformed Russia, but had little immediate impact on Vavilov's activities. After an uncertain start, dominated and limited by the imposition of communist dogma, Lenin introduced the New Economic Policy in 1921. Private business transactions were sanctioned and incentives offered to encourage foreign investment, to the dismay of many committed communists. Agriculture in particular benefited from this support, and exports of wheat provided substantial foreign revenue for several years. The prosperity of the *kulaks* following an especially abundant, bumper harvest in 1926 attracted the attention of hard-line communists after Lenin's death. They proceeded to replace the New Economic Policy they detested with the first of a notorious series of Five-Year Plans, placing the *kulaks* at the sharp end of reform. Private enterprise of any kind was penalized by prohibitive levels of taxation, and the Plan opened the way for nationalization of agricultural land throughout the Soviet Union by gathering together peasant holdings in state or collective farms. In 1930 Joseph Stalin denounced the *kulaks* as profiteers and saboteurs who put their selfish interests before the needs of their country. Their existence, he declared was incompatible with a socialist society. During the onslaught which followed tens of millions of families were dispossessed of their land and conscripted onto state farms (34). Many were summarily deported, very few ever to return.

The wholesale transition from a peasant economy to a cumbersome conglomerate of state-run collectives stretched the organizing abilities and skills of a hastily rearranged agricultural ministry beyond its competence. Within two years the fields were untended and uncropped. Throughout the Soviet Union, even in the most fertile and prosperous parts of the Ukraine, and the Kuban, a region on the Black Sea, millions died of starvation. The *kulaks'* cabins and barns had gone up in flames and with them their stores of grain; any that had escaped the destruction had been eaten later in desperate attempts to prolong life. Within a couple of terrible years the landraces of buckwheat, rye, oats, barley, sunflowers and wheat, which had been the mainspring of the country's agriculture, were wantonly destroyed.

The liquidation of the *kulaks* – and with them the agricultural base that sustained the Soviet Union – was a cataclysmic event, contrived by people who were neither aware of, nor cared about, the consequences of their dogmatic assumptions.

The collections of seed made first by Robert Regel and added to by Nikolai Vavilov and their colleagues working for the Bureau of Applied Botany (by then renamed the Research Institute of Plant Industry) preserved at least a part of the treasure from destruction. They also revealed the value of seed collections as a readily available reserve from which to replace plant resources lost through disasters of one kind or another – in other words as a means of conservation. Yet the collections had not been assembled with such a thing in mind. Few people at that time would have been thinking in those terms, and nobody before the events of the 1930s could have anticipated the fragility and vulnerability of Soviet agriculture's genetic base. The collections had actually been assembled to support plant breeding programmes aimed at producing new and more productive varieties. These were urgently needed by a country stretched across a vast area of the globe's surface, where crops were grown in innumerable different situations.

The next few years could have been the time when these collections, and the network of experimental stations set up to make use of them, came into their own. Perhaps they would have done so but for another dose of communist dogma. Vavilov, descended from a family of prosperous Moscow merchants, was inevitably an object of suspicion to the Bolsheviks who now ran the country. He had graduated from the Moscow Agricultural Institute before gaining a thorough scientific grounding at leading botanical research establishments in pre-Revolutionary Russia, another cause for suspicion, and, worse than that, had completed his studies (in 1913–14) by a visit to Britain where he had worked with Professor Bateson, one of the founders of the science of genetics (p. 173), at Cambridge. Vavilov had also established and maintained close and very friendly contacts with botanists, geneticists and agriculturalists here, there and seemingly everywhere. The combination of his antecedents, professionalism and liberal, outward-looking disposition was deeply unpalatable to the rulers of the Soviet Union. They were practically certain to lead to confrontation with Vavilov's communist rulers, and that they did not do so directly is a measure of the man's extraordinary integrity and force of character.

The crunch came over an ideological issue. The Bolsheviks were committed to the principle that nurture, rather than nature, moulded their citizens, and the country's new rulers insisted that society was malleable. Those who continued to believe that individuality and behaviour owed at least as much to the genes people had inherited were condemned as saboteurs and enemies of the state. The survival of the fittest and the inheritance of characters were dismissed as bourgeois concepts, especially dangerous because they promoted individuality at the expense of social conformity. Those who insisted on applying genetics to the breeding of superior, high-yielding, disease-resistant crops, such as Vavilov and his colleagues, were implicitly supporting innate class differences, condemned as elitist by the regime.

Unsurprisingly, there were people willing and able to pursue these ideas to their own advantage. One such was Trofim Lysenko, an ex-student of

Vavilov's with the advantage of a more acceptably unprivileged, evidently non-elitist, background. Lysenko discovered, perhaps almost by accident, that by dismissing Mendelian genetics he could make himself remarkably popular with Stalin. All this attention to breeding and genetics was quite unnecessary, he announced, because simply by exposing germinating seeds to a touch of frost he could conjure up strains of wheat capable of growing and cropping well in cold climates. Of even more value was the fact that future generations raised from these plants would also be frost-tolerant.

That was what the communist leaders of a country afflicted by a climate too cold to grow wheat over much of its area wanted to hear. Stalin supported him enthusiastically in person, and Lysenko was rapidly promoted as a rival to his former mentor. One bad thing led to another as the rising star exploited his position with further claims. He could treat spring-sown wheat to make it as productive as autumn-sown wheat; produce strains of wheat capable of growing well in infertile conditions or in arid parts of the country; confer wheat with resistance to pests and diseases. So it went on, eventually expanding into other agricultural crops and even the management of dairy cows.

Throughout the 1930s Vavilov was subjected to increasingly hostile criticism. The programme he directed was discredited on the grounds that he was wasting time going around collecting plants which were obviously useless, an activity tantamount, indeed, to sabotage. Harassed and forced into a corner by the communist state's support for Lysenko's opportunistic theories, Vavilov felt driven to declare in 1939, 'We shall go the pyre, we shall burn, but we shall not retreat from our convictions. I tell you, in all frankness, that I believed and still believe and insist on what I think is right…This is a fact, and to retreat from it simply because some occupying high posts desire it is impossible.'

Unfortunately for his future prospects, one of those occupying 'high posts' was Joseph Stalin. In August 1940 Nikolai Vavilov was arrested just before setting out on a plant collecting expedition to the Carpathian Mountains, and several of his colleagues were rounded up and imprisoned. Vavilov stood accused of sabotaging agricultural production, spying on behalf of the British government and conspiring against the communist social order.

31 In Albania in the late twentieth century humans and beasts, rather than tractors and combine harvesters, tilled the soil and gathered crops grown from home-saved seed.

32 Lupins are a diverse group of plants found at high altitudes. The majority of lupin species are found in the Andes. However, lupins familiar to gardeners have their wild origins in habitats such as this in North America.

33 The so-called Russell hybrids, so beloved by gardeners, were derived from North American lupin species by George Russell. From mixed collections of plants, Russell selected the different flower colours that he liked. Eventually purple lupins were transformed into a kaleidoscope of flower colour.

34 In the early days of the Soviet Union, agriculture and the people who worked on the land benefited greatly from the plant collecting and breeding efforts of Vavilov and the policies of the new communist government. However, by the 1930s, government attitudes changed and undid much of this work.

35 The American agronomist Norman Borlaug (1914–2009) is considered to be the father of the Green Revolution. He bred wheat cultivars to have short, strong stems that could support large, productive ears.

36 Monkombu Sambasivan Swaminathan (b. 1925) is an Indian agricultural scientist who is credited with introducing and developing high-yielding wheat cultivars into India. He is shown here examining plants that have been tissue cultured.

His trial concluded, inevitably, with the death sentence, subsequently commuted to life imprisonment. This was to prove only a short extension, as prolonged neglect aggravated by starvation led to his death in the southern Russian city of Saratov in January 1943. Vavilov's body was one of hundreds disposed of anonymously in a communal prison grave.

In the interval between Vavilov's arrest and his death, the German army had invaded the Soviet Union. The disorganized defence offered by a demoralized Red Army, torn apart by paranoiac purges, had been overwhelmed, and the invaders had reached the gates of Leningrad. Hundreds of thousands died of starvation during the siege that followed, including scientists and technicians working at the Research Institute of Plant Industry in the city. Some of the collections of seeds held there and elsewhere in the USSR had been moved east to relative security beyond the Ural Mountains, but much remained in the vaults. When the Germans finally retreated, an inventory of the collections of rice, corn, wheat and other seeds entrusted to the institute's care showed them, so we are told, to be still intact – an extraordinary achievement given the starvation inflicted by the siege.

The desperate times suited Lysenko's half-baked theories, and his pseudo-scientific charlatanry prospered while Stalin remained in power. Despite an inadequate grasp of scientific principles, and even because he denied the value of applying scientific methods to his experiments or statistical analyses to his results, Lysenko was progressively promoted. His approach provided the communists with a Marxist line on biology that could repudiate the premises of bourgeois science. 'Marxist biology' existed only as a woolly concept, but its adherents were sure they would know it when they saw it – and they saw it in Lysenko's propositions. An entry in a Soviet encyclopaedia, published in about 1950, provides a succinct and explicit insight into the situation: 'Gene is a mythical part of living structures which in reactionary theories like Mendelism-Veysmanism-Morganism determines heredity. Soviet scientists under the leadership of Academician Lysenko have proved scientifically that genes do not exist in nature.'

These heretical ideas gained ascendancy throughout the 1930s and retained a dominant position during the 1940s. They only began to decline after Stalin's death in 1953, when the ridiculous nature of Lysenko's assertions was once again exposed to criticism and fair comment. It took time, but the mood was changing. The seventy-fifth anniversary of the foundation of the All-Russian Scientific Research Institute of Plant Industry in 1968 was celebrated by adding the prefix N. I. Vavilov to the official title.

Lysenko's victory is a sad story, and a salutary example of the evils which follow when the state, or other controlling force, steps in and prohibits free discussion and the attainment of balanced perspectives. Experiments were done in ways that could not possibly produce meaningful results. They were interpreted by those who professed to believe their experiments would yield whatever results expediency and wishful thinking led them to expect. In the process, Lysenkoism demoted the seed from the means by which a plant expresses its genes to an incidental role as the starting point of a new plant. Everything that had been progressively revealed about seeds by Camerarius, Koelreuter and Sprengel, by the experiments of von Gärtner, Knight and Mendel, and most recently by the discovery of the principles of genetics during the early twentieth century was set aside as irrelevant.

The political drama, scientific turmoil and human conflict that engulfed all aspects of seed collection and plant breeding in the Soviet Union distract attention all too easily from what was going on elsewhere in the world. This included a collecting programme of at least equal ambition to Vavilov's, begun in 1898 by the USDA. International co-operation had been a hallmark of these activities from the early days when Robert Regel, then director of the Bureau of Applied Biology in St Petersburg, exchanged seeds and information with plant collectors in the USA. Vavilov maintained and extended such links until the disapproval of his political masters abruptly terminated them. As we shall see in the following chapter, however, they were renewed after the restoration of Vavilov's star under Nikita Khrushchev, providing a rare example of scientific cooperation across the political divide of the Iron Curtain.

The USDA's Plant Introduction Program is one of the longest established, continuously active agricultural programmes in both the USA and the world. It shared its one hundred and tenth anniversary with the sixtieth anniversaries of the founding of the National Small Grains Collection, responsible for maintaining and characterizing collections of wheat, barley, rye, oats and other cereals, and four regional plant introduction stations in different parts of the United States, where seeds and plants collected overseas were grown and evaluated. Last and by no means least, 2008 also marked the fiftieth anniversary of the construction of the National Seed Storage Laboratory at Colorado, the leading long-term storage facility for the seeds of economically important plants in the world. These anniversaries, all falling in a single year, encapsulate the history and indicate the scale of the USA's official efforts to introduce, trial and exploit seeds of plants of agricultural interest on behalf of plant breeders, farmers and gardeners, and to store them securely for the future.

Previously, plant introductions had been the preserve of nurserymen and private individuals. Many of them had been extremely active, and played important roles in adding to the diversity of different crops and varieties grown in the USA. Nurserymen, in particular those who had arrived as immigrants, used links with their homelands to arrange for the importation of plants with which they had previously been familiar. During the nineteenth century, at a time when nurserymen in Europe were enthusiastically exploring opportunities to breed new varieties (p. 56), US nurserymen seem to have been more inclined to import, rather than breed, new kinds. Possibly, because of the strong, fundamental Christian principles that many of them had been brought up to respect, they were more diffident about assuming the role of creators.

During the first years of the twentieth century David Fairchild was the man in charge of officially funded efforts to introduce new kinds of vegetables, fruits, grains, pharmaceutical and other potentially valuable economic plants to the USA. He was also a notable collector in his own right, credited with the introduction of some twenty thousand collections including avocados, nectarines, peaches, pawpaws, pistachios, various bamboos and soybeans. He was a public servant who performed his duties in the manner expected of him, and a great deal better than most. His sponsor, and companion on many of his collecting expeditions, was a wealthy friend called Barbour Lathrop. Fairchild was honoured by having the Fairchild Tropical Garden in Florida, one of the world's great collections of palms, named after him, but he was never involved in great conflicts between scientific integrity and opportunism like those which overwhelmed Vavilov. Lacking the quirky antecedents, charisma or downright peculiarities with which several of his collectors were endowed, Fairchild nevertheless established the tone and momentum of the USA's plant introduction programme.

Fairchild is remembered today less than he deserves to be, while all but a few of the collectors he employed are mere names in the USDA's archives. Those we know about tend to leave the impression that in its early days the USDA recruited its collectors from the same pool of incorrigible travellers that half a century later supplied tour leaders for the more adventurous travel companies. Two who qualify for their achievements are Frank Meyer and Jack Harlan, whom we will return to later. Another, who impressed as much by his idiosyncrasies as his introductions, was Joseph Rock.

Rock was born in Austria and brought up by his father, steward to the household of a Polish count and a dourly obsessive religious fanatic. The young Rock fled his home to escape his father's control, and after twenty years of travelling and living on his wits eventually became a US citizen. He acquired the post of Professor of Systematic Botany at Hawaii University, an appointment not entirely unconnected with his spurious claim to have graduated from the University of Vienna. By then, despite having no formal academic qualifications, Rock had become fluent in Hungarian, Italian, Greek, English, French, Chinese and Arabic, in addition to his native German. At the age of thirty-six he was struck again by wander-lust, and he applied to David Fairchild for a job with the USDA. Fairchild sent him off in search of the chaulmoogra tree, the source of an oil to treat leprosy.

The search took Rock through leech-infested jungles and along rivers thick with crocodiles. Yet he travelled with his own servants whose duty it was to prepare a canvas bath at the end of each day's trek, and to lay out a freshly laundered white shirt, jacket and tie while their master was relaxing in the water. Refreshed and suitably dressed, Rock would sit down at table (laid in a civilized fashion with a linen cloth) to eat a European-style dinner with silver cutlery, dished up on fine crockery by a native cook trained in his master's ways. Clearly his upbringing in the stately home of a Polish count had left Rock with a taste for life's refinements.

Despite his eccentricities, Rock eventually found the chaulmoogra trees in the jungles on the borders of Assam and Burma and, so his account implies, snatched the ripe fruits from the jaws of the waiting wild swine and bears. A tigress even tore down a flimsy hut in the village where he was lodging, and dragged out and ate two women sheltering inside. He returned undaunted to the USA, however, with the seeds needed to set up plantations for the commercial production of the oil. Unfortunately, the oil turned out to be less effective than had been claimed.

David Fairchild sent Rock off again – this time to China, where he would find his spiritual home. Rock became an aficionado of all things Chinese, making many collections, notably of rhododendrons, in the unsettled, bandit-infested country in the southwest of the country, where Moslem and Tibetan inhabitants of the borders between China and Tibet were in constant conflict. Eventually this introverted, isolated man with his little pretensions and inability to relate to his fellows found a resting point and solace of a kind in western China – only to be slung out by the communist authorities after Mao Zedong's victory in 1949. Rock returned to Hawaii, where he died in 1962.

In 1889 a fourteen-year-old lad called Franz Meijer started work as a garden boy at the Botanical Gardens in Amsterdam. A few years later he was put in charge of the experimental garden where his abilities attracted the attention of Hugo de Vries, who took him under his wing and made arrangements for him to attend some of the courses in botany provided by the university. Armed with these qualifications, he set off for the USA, via England. Franz Meijer found a job with the USDA and became Frank Meyer, working for a year as a bacteriologist before transferring to the Plant Introduction Station in Santa Ana, California. This led to a spell of collecting in Mexico, which in turn led to him being asked by David Fairchild to undertake a collecting expedition to China.

Meyer was born and bred in a small, flat country where walking was as good a way as any of getting from place to place, and he never lost the habit. He arrived in Peking in 1905 with an itinerary that included a vast area of northern China, including a corner of Mongolia, Manchuria, Korea and down to Shanghai. Meyer duly set off on a walk of more than a thousand miles that was to take him two and a half years. Along the way he collected seeds of numerous fruit trees and a variety of other, mostly woody, trees and shrubs, as well as a kind of spinach that attracted his attention because of its large, succulent leaves. Meyer sent seeds back to the USA, where it was discovered to be more than just a spinach plant with large leaves. Meyer's plant possessed genes resistant to blight and wilt, which were used by USDA's plant breeders to produce new disease-resistant varieties.

During the next ten years, Meyer carried out three further seed collecting expeditions. Two were to China, and one lengthy excursion started in the Crimea, extending along the southern borders of Russia before moving ever east through Armenia, Azerbaijan, Chinese Turkestan and the Russian-Mongolian border. Here civil unrest made it too dangerous to move on into China as intended, so Meyer retraced his steps across the roof of the world. He eventually sailed from Southampton bound for the USA on 9 April 1912 – one day ahead of the *Titanic*. While in the vicinity of Saratov, on the Volga in southern Russia, he collected seeds of a crown vetch. These seeds were later used to produce the variety Emerald, widely grown as fodder for animals in Iowa.

When Meyer returned to China the following year, he was a man with a mission – to discover whether chestnut blight had originated in Asia, and if so to find sources of resistance to the disease. Before 1900, the American chestnut had dominated forests of the eastern seaboard of the USA. However, by 1910 these forests had been decimated by the introduced fungal disease chestnut blight. There was no natural source of resistance in native American chestnut populations. Meyer was able to prove that chestnut blight had come from China, but the short-lived nature of chestnut seeds frustrated endeavours to introduce resistant strains. During the next two years he travelled from one side of the country to the other, making collections of seeds and scions of fruit trees. While on the borders of Tibet, the trials and difficulties of his travels got the better of Meyer, who, during a disagreement with his interpreter and a coolie labourer, threw the pair down a flight of stairs. Brought up before the local magistrate, he might have ended up in a Chinese prison had he not found an ally to speak up for him in the shape of the British plant collector Reginald Farrer, who was also plant hunting in the region.

Meyer returned finally to China in 1916 after staying only a few months in the USA, this time to collect seeds of wild pears resistant to fireblight. He did so, and travelled on up the Yangtse river. On his return, China was in such an unsettled state that it was too dangerous for him to travel beyond Ichang; he holed up there for the winter until June 1918, when he boarded a ship bound for Shanghai and home. The voyage ended tragically almost before it had started. On the first evening he disappeared, and his body was later recovered from the river, although whether he fell from the boat, jumped or was pushed nobody will ever know. Frank Meyer's enduring memorial is the soybean. Before he set off on his first walking tour of China, US farmers grew only eight different kinds, none of them very satisfactory. His collections added more than forty new varieties, the foundation of today's soybean industry in the USA.

A year before Meyer's untimely death, a man was born who would take up the mantle of Nikolai Vavilov. Jack Harlan, second son of Harry Harlan who worked on barley for the USDA, became one of the leaders of international efforts to prevent the loss of the genetic resources upon which plant breeders depended. He was to be one of hundreds of unsung plant breeders, collectors, seed bankers, agronomists and extension workers who contributed to the largest enterprise in the world concerned with gathering together and using plant resources. During seed collecting expeditions to Europe, Asia, South America and Africa, his father became friendly with many other plant collectors; one was Nikolai Vavilov, whom Harlan welcomed as a visitor to his home in 1932 during an international congress on genetics. His son Jack, then aged fifteen, met Vavilov during the visit, and the meeting made such an impression that the boy determined to become a plant collector himself. Plans for Jack to go to Leningrad after leaving school were aborted when Vavilov's relations with the Soviet authorities deteriorated, and he followed a more conventional academic course instead. After graduating from George Washington University, Jack moved on to a post-graduate doctoral degree at University of California, Berkeley.

Jack Harlan set off in his father's footsteps when he found a job with the Department of Agriculture in 1942, breeding forage crops and improving the grazing quality of rangelands in Oklahoma. His interest in parallels between these practical activities and the improvement of cereal grasses during the early stages of plant domestication marked the start of a long career devoted to the study of the origins of crop plants, the people who domesticated them and the use and conservation of plant resources. Harlan left the Department of Agriculture to become a full-time academic, first at the University of Oklahoma and then at the University of Illinois. Here he co-founded the influential Crop Evolution Laboratory, in 1967, with Johannes de Wet.

Harlan was now able to indulge his commitment to plant collecting. He carried out more than forty expeditions to countries all over the world, observing the conditions in which plants grew in cultivation as well as in the wild, and over and above that playing a leading role in a series of archaeo-botanical studies on the origins and domestication of crop plants. The USDA sponsored much of this work, and was well repaid by the extraordinary number, diversity and quality of the collections Harlan made on their behalf, including numerous forest trees, fruits and ornamental plants. However, in accordance with his main interests, wheat, barley, maize, clovers, medicks, trefoils and other forage legumes and grasses were the focus of his main efforts, and many of these would play their part in official plant breeding programmes.

One of his most notable collections, with a Turkish colleague called Osman Tosun, was from a wheat field in a place called Fakiyan Semdinli in eastern Turkey. Not even the farmer in whose field the plants were growing could have regarded them with pride. They grew tall and gangly, were not very hardy and looked miserably unthrifty, while their thin, weakly constructed stems were unable to support the burden even of their inadequately filled ears. Furthermore, the small yields of grain the plants provided had poor baking qualities. The collection was duly registered in the USDA's seed bank as PI 178383, where it remained, disregarded and unappreciated for some fifteen years.

In the early 1960s a severe outbreak of stripe rust began to decimate wheat crops in America's northwestern states. Hundreds of collections were tested for resistance to the troublesome strain of stripe rust, and PI 178383 came up trumps. Further tests revealed that it was also the carrier of genes providing resistance to three other strains of stripe rust, thirty-five races of common bunt, ten races of dwarf bunt and tolerance of flag smut and snow mould. The miserable-looking runt had turned about to be a veritable fungicidal bomb. Plant breeders became all too eager to incorporate its genes in their new introductions and varieties bred from it dominated the wheat-growing scene in the northwestern states. Forty years after PI 178383's triumph a twist was added to the tale when another USDA collector went to Fakiyan Semdinli to see if the plant had survived there. Unsurprisingly, it had not, but he did discover that it was probably not Turkish wheat at all. The farmers in whose fields it had grown were immigrants from Iraqi Turkestan, and had probably brought the seed with them from their old home.

Jack Harlan's exceptionally extensive practical experience of collecting and his observations of cultivated plants throughout the world enabled him to review Vavilov's theories, and evaluate how they had stood the test of time since they were proposed in 1926. Although Harlan preferred the term 'centres of diversity' to Vavilov's use of 'centres of origin', and refined some of Vavilov's other conclusions, he found himself in general agreement – most significantly with the crucial importance of a few locations as the crucibles in which much of the crop diversity on which plant breeders depended had been generated (p. 116). In the latter part of his life Harlan became one of the major players in the campaign to conserve these irreplaceable resources.

Unlike Russia and the United States, the British government made no centralized arrangements for the collection and exploitation of the genetic resources of crop plants. There were other initiatives during the twentieth century, such as government-financed establishments for breeding wheat, barley and other cereals at the Plant Breeding Institute at Cambridge, and for grasses and forage crops at the Grassland Research Institute near Maidenhead and the Welsh Plant Breeding Institute at Aberystwyth. However, from the eighteenth century, Britain, as had other European imperial powers, had developed a very strong and worldwide interest in plant collecting – based almost exclusively on the initiative and activities of private individuals.

Joseph Banks's voyage with James Cook had opened his eyes to the diversity and splendour of the wildflowers in distant parts of the world.

When Banks became the *de facto* director of what would eventually become the Royal Botanic Gardens at Kew, he commissioned a succession of young gardeners as plant collectors – an example followed up by the officers of the Royal Horticultural Society, notably the assistant secretary John Lindley. He in turn was responsible for sending one of the most successful and eminent of all plant collectors, Robert Fortune, to introduce the newly revealed treasures of the gardens of China to British gardeners. Fortune is perhaps best known for his introduction of Chinese tea plants to India. However, during the 1840s to 1860s he introduced plants such as kiwi fruit, Chinese red bud and forsythia to British gardens.

Seeds played lesser parts than living plants in these introductions until Joseph Hooker, son of William Hooker (now director of the Royal Botanic Gardens, Kew) set off to collect plants in Sikkim in 1848. He returned, after a series of harrowing misadventures with numerous collections of seeds, with rhododendrons. Hooker's Sikkim rhododendrons were a revelation to gardeners in Britain and sparked an interest in the flora of the eastern Himalayas, the borders of India, China and Burma and western China, which inspired a series of plant collecting expeditions to those parts of the world. These highly productive expeditions provided the wherewithal for the development of the woodland gardens that played such crucial roles in the development of gardening in Britain during the twentieth century, with an emphasis on primulas, witch hazel and anemones.

Rhododendrons, in particular, make life easy for seed collectors. Not only are they extremely prolific, but their seeds forgive being collected while still green more than most plants, greatly extending the season when worthwhile collections can be made. Seeds also provided much more conveniently transportable and more resilient collections of primulas, gentians, meconopsis and other favoured plants than actively growing specimens. However, mature seeds are available for only a short season, and, given the uncertain conditions with which collectors had to contend, being in a particular place at a certain, predetermined time was a major difficulty.

George Forest, a sturdily built, dour Scot, was employed at the Royal Botanic Garden, Edinburgh, and sponsored by the Liverpool cotton broker Arthur Bulley and, later, John Charles Williams of Caerhays Castle in

Cornwall. He epitomized the successful seed collector, and the quantities of seed he collected swamped the resources of the gardeners amongst whom it was distributed. Forest's secret was to employ numerous native collectors, whom he recruited from the local hill tribes. Trained to identify plants correctly and make successful collections of seed in due time, they could be relied upon to scour the countryside on Forest's behalf. Between 1904 and 1931 he collected plants of the mountain flora of Yunnan and neighbouring parts of Burma and southeast Tibet on seven expeditions with phenomenal success.

The numbers of collections he made were truly staggering, including more than five thousand three hundred collections of rhododendrons, and more than one hundred and fifty species and sub-species of primulas, in addition to comprehensive collections of dozens of other genera. At the conclusion of his final collecting expedition he wrote home, 'of seed such an abundance, that I scarcely know where to commence, nearly everything I wished for and that means a lot. Primulas in profusion, seed of some of them as much as 3–5 lb, same with *Meconopsis, Nomocharis, Lilium*, as well as bulbs of the latter. When all are dealt with and packed I expect to have nearly if not more than two mule-loads of good cleaned seed, representing some 400–500 species, and a mule load means 130–150 lbs. That is something like 300 lb of seed. If all goes well I shall have made a rather glorious and satisfactory finish to all my past years of labour.' Forest died a week or two later from a heart attack in his fifty-ninth year, while still in the field – a fulfilled, if not a happy, man.

By the early twentieth century, plant breeding by private companies as well as the activities of the USDA – fuelled by the efforts of collectors in many countries, and maintained by a small army of geneticists, agronomists and extension workers liaising with farmers throughout the country – had transformed agricultural production. It had also achieved great success in breeding cereals and other crops with disease resistance, hardiness, drought tolerance, productivity and other qualities needed to sustain high yields.

During the 1940s the US Government decided to use this plant breeding expertise for political purposes. They turned first to Mexico, underpinned by funds provided principally by the Rockefeller and Ford Foundations, with the laudable intention of improving the lot of the rural poor – and in the hope that by doing so they would increase social stability and reduce communism's appeal. The Cooperative Wheat Research and Production Program was set up under the auspices of the rather cryptically named Office of Special Studies to breed improved varieties of wheat and maize.

The man in charge of the project was Norman Borlaug (35), son of an Iowan farmer of Norwegian origin. He proved to be a man who combined a remarkable clarity of vision with a ruthless ability to pursue a commitment. Borlaug had a strong preference for practical, clearly defined targets, together with a certain disdain for academic niceties. After completing a degree in forestry at the University of Minnesota, he worked with the US Forestry Service before returning to college to study plant pathology as a postgraduate. He then became head of research projects on bactericides and fungicides at the Du Pont de Nemours Foundation at Wilmington in Delaware. Two years later, in 1944, Borlaug was chosen to take up a post as geneticist and plant pathologist in charge of the plant breeding programme in Mexico.

Despite qualifications which do not immediately strike one as ideally suited to the needs of a plant breeding establishment, Borlaug was outstandingly successful as co-ordinator. The project included several of his colleagues: the maize breeder Edwin Wellhausen, John Niederhauser, who specialized in potatoes, and the agronomist William Colwell. Soon after starting the project, Borlaug had a thought. Surely it should be possible to grow two crops of wheat a year: one during the summer in a mountainous part of central Mexico, the other during the winter on the tropical coastal plain. The idea received a frosty reception from colleagues trained as orthodox agronomists. They assured him it would not work because wheat seeds had to have a rest after being harvested, that is, to after-ripen, before they were able to germinate. Borlaug had little use for orthodox theory and, after considerable opposition, was given approval to try his idea out – with complete success. Not only was he able to halve the time it took to breed a new variety, he also benefited from an entirely unexpected bonus. Seeds sown in central Mexico grew during the long days of summer; those on the coastal plain during the short days of winter. Only varieties that performed well in both situations made it through to the final selection, with the result that Mexican wheat, as his productions came to be called, had no affiliations to day length; it could be grown anywhere in the world where other climatic conditions allowed.

Nevertheless, Borlaug's orthodox colleagues were right to be sceptical. A less satisfactory consequence of his cunning dodge to accelerate his breeding programme was a reduction in the length of the after-ripening period, and an increased tendency for the grain produced by his varieties to germinate soon after it matured. In wet years this can lead to problems and heavy wheat losses when grains in the ears germinate before they can be harvested.

During the 1950s Mexico became self-sufficient in wheat and by the 1960s the country was producing an exportable surplus. The key to success was the introduction of dwarf forms of wheat, originally bred in Japan, which made highly effective use of nutrients produced by photosynthesis. Eighty per cent of the energy obtained from sunlight by plants of unim-proved landraces goes into the production of straw, and only 20 per cent to the seeds. Modern, dwarf varieties of wheat, rice and maize transfer about 50 per cent of the energy to their seeds. What is saved on the straw goes into the seed, which (unless you happen to be a professional thatcher) must be a good thing. The productivity of maize plants was similarly increased, partly by reducing the amount of growth that went into leaves and stems, but mainly by producing plants capable of growing closer together without reductions in yield. This was done by breeding plants with less spreading foliage and a more upright stance, which reduced shading between neighbouring plants.

Later the Office of Special Studies became the International Maize and Wheat Improvement Center. Known from 1943 as CIMMYT, it was an autonomous research institute with a board of governors and international staff, who enjoyed many of the privileges normally accorded to diplomats. Borlaug was made director of the International Wheat Improvement Program, with responsibility for organizing the radical transformation of methods of agriculture in many countries, to be dubbed the 'Green Revolution'.

The success of the enterprise in Mexico confirmed the US government's hopes that political advantages might accrue from exporting the promise of increased crop yields to developing countries. The programme had two heads. Plant breeders saw it as a humanitarian cause to increase the productivity of crops in parts of the world where this made a crucial contribution to the welfare, health and quality of life of the people, especially farmers – many of whom were poor and working at little more than a subsistence level. The government saw it as a political weapon in the Cold War – one that provided an antidote to the egalitarian blandishments of communism by reducing poverty and deprivation.

One of the first to receive the benefits was India. This was due in large part to the inspirational leadership of an Indian agronomist, Monkombu Sambasivan Swaminathan (36) – a visionary who, early in his life, dedicated himself to eliminating hunger and poverty in Asia. After graduating in India, Swaminathan studied in Holland, then as a postgraduate in England, before taking up employment in the USA, working on research into the genetics of potatoes and the breeding of new varieties. He returned to his native country, convinced his skills were more needed in Asia than in North America. While working at the Indian Agricultural Research Institute, Swaminathan started crossing local varieties of wheat with Mexican varieties produced at CIMMYT, and with other dwarf strains imported directly from Japan.

During the 1960s the Indian government, backed by the Ford Foundation, imported huge quantities of wheat seed from Mexico to bolster its own plant improvement programmes under the direction of Swaminathan and his colleagues. The programme was spearheaded by an array of two thousand model farms to show farmers how to use fertilizers and pesticide sprays in the most effective ways. Swaminathan was a man widely respected for his belief in the benefits of using environmentally sustainable methods of agriculture and the need to safeguard biodiversity – making his advocacy of the Green Revolution's technical approach to solving problems particularly significant.

Whatever reservations Swaminathan may have had about the methods, the results were astonishing. The new kinds of wheat were introduced to

India in the mid-1960s, and within two years wheat production in India had risen from ten to eighteen million tons; by 2004 it had reached eighty million tonnes annually. The spectre of famine, once a seemingly inevitable part of existence in many parts of the country, began to fade into memory.

What had been achieved with wheat in Mexico was applied to the Philippines in 1960, when the Rockefeller and Ford Foundations established the International Rice Research Institute (IRRI), on similar lines to CIMMYT. The role of IRRI was to make collections of varieties grown throughout southeast Asia, breed new and improved varieties and distribute them to countries in the region. One of the principal objectives was to produce dwarf, high yielding rice varieties based on those bred in Japan, capable of growing under tropical conditions. The outcome, the so-called 'miracle rice', did for tropical and sub-tropical farmers what 'dwarf Mexican wheat' did for those who farmed in cooler parts of the world.

High yielding varieties were the hallmark of the Green Revolution, but its effects went much further than simply replacing unproductive landraces with pedigree strains of seed. The secret behind the high yields was the ability of these new varieties to make more effective use of nitrogen; to support their needs, the plants had to be supplied with artificial fertilizers at levels far in excess of those previously used in peasant agricultural systems. Similarly, achieving high yields depended on freedom from the depredations of insects or fungal infections, and that called for the application of pesticides of various kinds. Finally, irrigation was now essential in drought-prone parts of the world.

The programme was highly successful – not least because, at Borlaug's insistence, it included a strong commitment to training agronomists and extension workers in the countries to which it was applied to run things for themselves. Yields of wheat, rice and maize in many developing countries more than doubled between 1961 and 1985 as more and more farmers replaced their landraces with pedigree varieties and were taught to apply new agricultural techniques. Almost three-quarters of the rice grown in non-communist Asia by 1990 was grown from seeds bred as part of the programme; as was half the wheat in Latin America and Africa, and nearly 70 per cent of maize worldwide. Nor was that all. In addition to the major cereals, improved varieties of barley, sorghum, millet, cassava, beans and many other crops made their contributions to the increased productivity of farms and smallholdings across the world.

A remarkable aspect of the worldwide Green Revolution was its timing in relation to similar improvements in agricultural productivity in the USA. Barely twenty-five years before US government-sponsored organizations began to export the products of improved plant breeding and novel approaches to the use of fertilizers, American farmers – notably in the corn belt of the Midwest and the wheat growing areas of the north – were being dragged (often with considerable reluctance) along the same path. Farmers emphasized caution and staying lean as the keys to survival during the depression years of the 1930s and into the 1940s, with their combination of agricultural surpluses and low prices. They were more inclined to save money by purchasing seeds cheaply and not using fertilizers than to increase productivity by investing in the (comparatively very expensive) seeds of the newly developed F_1 hybrid maize varieties, along with bags of artificial fertilizers which their forebears had used sparingly or not at all. Improvements in the economic situation within the USA, and shortages of food in other parts of the world, were needed to persuade US farmers that greater productivity was the way to make greater profits, and that investment in more productive varieties and technological advances was necessary. By the 1970s cereal seeds were being sown, and even the first fruits gathered, of an almost precisely similar transformation of agricultural methods in Mexico, India and other entrants into the new technological world.

After two centuries we appeared to have the answer to Malthus's gloomy prediction that human increase was bound to exceed the world's capacity to feed us. In 1970 Norman Borlaug received the Nobel Peace Prize in recognition of his services to the alleviation of poverty and hunger through the successes of the Green Revolution. However, unintentional consequences of the Green Revolution were to pose a serious threat to future human food supplies.

CHAPTER EIGHT

Banking on Seeds

Even as Norman Borlaug acknowledged the plaudits of the eminent audience attending the Nobel Peace Prize ceremony, metaphorical brickbats were heading in his direction. To many he was a hero. Those who believed that feeding the world depended on deploying technological advances in plant breeding – and the liberal use of fertilizers and pesticides – saw him as the saviour of the world. To others he was the man responsible for the destruction of the long established methods of farming on which future survival depended. In one corner were those who measured success by productivity and income, obtained from monocultures of high yielding, genetically uniform varieties, with ruthless control of as many aspects of the environment as possible. In the other were the traditionalists, committed to maintaining and enhancing genetic diversity. They put their faith in semi-natural ecosystems, and were prepared to modify conditions to suit the crops they grew – provided that they remained firmly based on natural balances between plants and their environment.

The Green Revolution increased cereal yields, but in doing so it inevitably transformed, and very often destroyed, long established methods of farming. Subsistence farming was replaced by high input, high output systems in which cash was essential to pay for seeds (previously home-saved), as well as fertilizers and pesticides and, very often, irrigation. Smallholders and peasants, unused to a cash economy and often drastically under-financed, could sink into debt; if they failed to repay what they owed, their land was forfeited and absorbed into larger estates. The advantages of mechanization also gave more wealthy farmers a huge advantage. Those unable to compete in the new world of farming were progressively dispossessed; many migrated to the cities in search of work, where they were only too likely to end up in urban slums, living hand to mouth. In many places the increased levels of fertilizers and pesticides resulted in the destruction of plants and animals and the pollution of rivers, lakes and other waterways. Irrigation brought with it the familiar spectres of salinization and falling water tables, especially in the dry, desert regions of the Middle East.

To some, these changes are the price of progress – a price that must be paid to lift world agriculture up to the productivity levels required to feed expanding populations. Others regard them as wantonly destructive of people's ways of life and systems of agriculture which, with more patience and a great deal more understanding, could be modified to transform lives and create more productive systems. Many would have preferred Green Evolution to Revolution, but whether a gentler process would have been capable of delivering the very impressive results of the more drastic approach is a moot point. This is a debate in which there are many good intentions on both sides. Nevertheless, the contention that gentler, more natural processes could have achieved anything like the yield increases reported over the years for countries touched by the Green Revolution, with its hard-nosed application of technology, appears optimistic in the extreme. Nor should we ignore lessons from elsewhere in the world of the pitfalls that beset attempts to impose dogmatic assumptions on agricultural systems. Even as Norman Borlaug was guiding the introduction of new and improved methods of agriculture to much of the world, in China, Mao Zedong's ideologically inspired Great Leap Forward created the greatest famine in human history between 1959 and 1961, during which tens of millions of people perished. The Green Revolution may be open to criticism, but we should give thanks that it was well organized, achieved its objectives of raising productivity (sometimes to an astonishing extent) and, in general, transformed the ways of life of its participants for the better.

The pros and cons of the Green Revolution are, and will remain, a matter for debate. However, few would disagree that initial hopes of a final answer to Malthus's gloomy thesis of 1798 – in which he claimed that human increase was bound to exceed the world's capacity to feed us – have had to be discarded. What cannot be doubted is the Green Revolution's impact on the reservoir of genetic diversity, built up over thousands of years by unsophisticated peasant farmers. The greater part of this genetic diversity was generated in a very few comparatively limited parts of the world – regions which, until very recently, seemed unlikely to be touched by modern agriculture. The Green Revolution revealed this reassuring view to be pie-in-the-sky.

Modern farming is based on high yielding strains of crops, bred to take advantage of good growing conditions, abundant water and nutrients, and freedom from pests and diseases. Since yield is a largely quantitative character, comparisons between varieties are likely to identify one or very few capable of producing the highest yield, and inevitably lead to the repetitive use of particular gene combinations of proven value at the expense of overall diversity. Favoured pedigree strains of particular crops tend to be related, and hence share many genes in common. Their genetic base is thus even more restricted than a head count of the numbers of varieties would suggest. Every variety of wheat growing over an enormous area might lack genes conferring drought tolerance, leading to calamitous failures when rains fail. Every kind of maize might share diminished resistance to a fungal disease which, once established, could spread unchecked from field to field. Insect pests that found congenial homes and agreeable fodder on one rice plant might find similarly encouraging conditions wherever they alighted.

For example, during the search for parents from which to produce F_1 maize hybrids a male sterile plant was found. This was considered a boon for breeders, who wanted to be certain that their seeds were the result of a cross between two different plants and not the result of self-pollination. Descendants of this plant were so widely used for the production of hybrid maize seed during the 1950s and 1960s that it became the ancestor of almost 80 per cent of the maize grown in the American Corn Belt. Similarly, genes derived from the wheat PI 178383 were, in the 1960s and 70s, overwhelmingly predominant in the wheat producing areas of the northwestern USA. For better, the resistance they conferred to stripe rust disease contributed to increased wheat yields; for worse, genetic uniformity made them all equally susceptible to any other affliction that might arise.

Pedigree strains of cereals have ousted landraces and comparatively unproductive, archaic varieties in one country after another. This success is now under threat, and paradoxically the threat is all the greater because the programme has been so successful. Before the mid-twentieth century if some plants in a wheat field succumbed to a fungal disease they were only a few among many, and some at least of their companions would be resistant to the disease. When severe droughts occurred, part of the crop would be lost, but individuals or varieties with greater drought tolerance survived to produce a subsistence crop. Now the genetic homogeneity of crops being grown over enormous tracts of countryside is such that a single disease or the climatic vagaries of a single season threatens disaster.

In some developing countries, the initial high yields of improved varieties have not been maintained. After a few years, particularly if adverse weather conditions occur, yields have fallen to the point where claims have been made that they scarcely exceed, and may even be below, those of the traditional crops they replaced. Sometimes these yields can be traced to the evolution of new strains of pathogens or increased levels of infestation by pests. A wry belief among farmers in Pakistan links the introduction of each new miracle variety of cereal to the imminent arrival of a new miracle locust.

Losses from drought, insects or fungi are nothing new. They have been a perennial hazard throughout the history of farming and gardening. But whereas in the past the consequences were limited by the presence of resistant or tolerant strains within the heterogeneous genetic composition of the crops, losses today may be serious, widespread and even all-embracing. Plant breeders have attempted to counteract this problem through a technique known as multi-line breeding, in which mixtures of varieties with different genotypes but similar characteristics are grown together, simulating in a very small way the diversity of the old landraces. This can reduce losses due to the appearance of a new strain of a pathogenic fungus, but new strains to which there is little or no resistance continue to arise.

Sometimes it has been possible to find disease resistance or other necessary qualities within the range of breeders' line or varieties available in agriculturally developed countries. When this has been impossible, plant breeders have had to fall back on the vast array of landraces maintained by peasant agriculture. Jack Harlan's ineffably weak-strawed, scruffy wheat plant (p. 192) is an example often quoted. Others, which might equally have hit the headlines, include a barley sample of little or no obvious agronomic appeal collected in Ethiopia. Cereals have moderately high protein levels, but their value as animal feeds tend to be limited by less than adequate concentrations of the amino acid lysine, one of the building blocks of protein. Some long established varieties of barley have desirably high levels of lysine, but are invariably low in protein. Barley that combined high protein with high lysine was a tantalizing but apparently unattainable prize, until this combination of characters was identified in a collection of barley made in Ethiopia.

Cake made from the crushed residues of rape seeds, after the oil has been expressed, is a valuable animal feed. In the 1970s high levels of toxic sulphur compounds limited the amount that could be safely fed to animals, until a variety of rape, grown in Poland and considered of limited value for its oil, turned out to have unusually low sulphur levels. Varieties combining both desirable qualities have now been bred by crossing low sulphur Polish rapes with high oil-producing strains.

Tomato growers plagued with fusarium wilt have even had to resort to grafting each plant onto a fusarium-tolerant rootstock in an effort to produce worthwhile crops – a process which is completely uneconomic on a field scale. The wild redcurrant tomato produces masses of tiny, currant-sized fruits that have no commercial value. However, a collection of redcurrant tomato, made in 1929, brushed aside infections of fusarium. Following crosses with its domesticated relative, its genes now contribute fusarium resistance to numerous commercial tomatoes (37). Other wild tomato species have also been drawn into crucial plant breeding programmes. A runtish little plant growing within a few yards of the Pacific Ocean in the Galapagos Islands turned out to be tolerant of saline soils. Seeds from another, collected among rocks and cacti in Peru, which most of us would have passed by without a second glance, have bequeathed drought tolerance to its descendants.

When bacterial wilt started causing unacceptable losses among field-grown beans, no fewer than one thousand six hundred collections, stashed away in the seed bank of the US National Program for the Conservation of Crop Germplasm at Fort Collins in Colorado, were trawled in order to find one resistant to the disease. This one collection was all that was needed to breed the resistant varieties now widely grown throughout North America. However, this has had the inevitable result that almost all the field-grown beans across an enormous tract of the USA now share a dangerously narrow genetic base.

The search for sources of resistance is a never-ending contest between the ability of fungi to mutate into new strains with unchallenged virulence, and the plant breeder's ingenuity in finding and using resistant genes. In 1999, fields of wheat in Uganda suffered from a particularly virulent strain of black stem rust (38). This is a particularly serious disorder because it attacks the whole plant and can completely destroy a crop, unlike diseases confined to the foliage that reduce yields but are unlikely to be devastating. Year by year infestations of stem rust have spread north through Kenya and Ethiopia; by early 2008 they were moving east through Yemen and towards Iran. In a year or two it seems highly likely that stem rust will have spread to the Punjab, the bread basket of Pakistan and India, and north into central Asia and Kazakhstan. The countries likely to be affected produce some 25 per cent of the world's wheat crop, which is the staff of life for hundreds of millions of people. At present there is nothing available to replace the disease-susceptible wheat varieties widely grown across the region. Several promising sources of resistance have been identified, but it will be at least five years before acceptable commercial varieties can be selected and sufficient seed produced to cope with the situation. The first appearance of this disease happened to strike Uganda, and it was some years before it threatened major wheat-growing regions of the world. The mutation which gave rise to it could have occurred just as easily in the Punjab itself, in the wheat belt of Western Australia or in the prairie states on the borders of Canada and the USA. The potential consequences of such a disease arising in the heart of one of the great wheat-growing areas of the world, with no time to set in train breeding programmes to counter its effects, are almost unimaginably dire.

These examples, and there are many others, demonstrate that although landraces and obsolete varieties have been superseded by the efforts of the plant breeders, they remain valuable as the sources of qualities which can be the salvation of an industry. Nevertheless, their decline proceeded almost unnoticed until well into the postwar era. Some individuals had commented at the end of the nineteenth century, but it was Vavilov in the 1920s and Jack Harlan in the 1930s who drew attention to the significance of their loss.

Pedigree seed strains, as might be expected, have long been the rule in western Europe and North America. Yet the old landraces were still preserved until quite recently in some parts of Europe, such as southern Italy, Greece, and the east – only to be lost as the Green Revolution swelled like a tide throughout the world. The durum wheats grown in Greece since time immemorial were reduced to a scant shadow of their former abundance within a few years. When concerted efforts were first made in 1967 to collect seeds of the landraces grown by farmers in Turkey, local strains possessing all sorts of different qualities were readily found. Five years later the fields were filled with imported kinds of wheat and barley, and only the most ancient, most conservative and obdurate peasants continued to grow traditional varieties. Similarly, in Iraq and Pakistan, ancient areas of cultivation stretching back to the dawn of cultivation (and once notable for the diversity of their crops) became filled with fields of pedigree cereals with high yields, but low genetic diversity.

Pedigree varieties can produce larger crops of higher quality grain than the landraces ever did in even the most favourable years. The agricultural benefits of these introductions have been enormous, but the loss of the resources needed by plant breeders to maintain the flow of new introductions has been profound.

The speed with which long established methods of agriculture and their complement of different kinds of wheat, barley, pulses, oats and rye surprised everybody. A survey of the status of landraces and primitive strains of cereals in Afghanistan during the 1960s concluded reassuringly that they were considerable and not in serious danger of being diminished. It maintained that they could be relied on to act as a reservoir of valuable breeding material for the foreseeable future. The printer's ink on the report was scarcely dry before severe drought, crop failure and famine in many parts of the country changed the situation forever. Farmers and their families had no option but to eat their seed corn in order to survive, and when the drought broke and they sowed their seeds to produce another crop, the seeds they sowed were of imported pedigree varieties. The fact that seeds were available averted the most severe consequences of the famine. Nevertheless, within two years conservationists had witnessed the wholesale destruction of resources they believed to be secure, without the slightest chance of being able to do anything to save them. Although the examples cited have been concerned predominantly with cereals in Europe and western Asia, similar cases affect agricultural production in many parts of the world.

By the 1960s the problem had been revealed and could no longer be ignored, but what could be done about it? There are several ways genetic diversity can be safeguarded. Biologically the most attractive is to continue to maintain landraces in the conditions in which they evolved by following unsophisticated peasant agricultural practices. Genetic diversity is then maintained *in situ*, in dynamic association with closely related weeds and semi-domesticated plants that have contributed to the development of today's crop plants. Alternatively we can collect seeds of representative samples from the fields where they can still be found, and conserve them elsewhere, generally referred to as *ex situ* conservation. The great collections built up during the first part of the twentieth century in the Soviet Union, at what was then the All Union Institute of Agricultural Sciences in Leningrad, and in the course of the USDA's extensive seed and plant collecting expeditions, consisted of collections made in this way (pp. 116 and 191).

The attractions of *in situ* conservation, following the age-old methods of peasant agriculture, are negated by almost insuperable practical problems, and almost invariably would be impossible to do on a scale capable of producing a worthwhile result. Any suggestion that a large enough number of farmers might be persuaded to revert permanently to old-fashioned, primitive forms of cultivation for reasons which would be at best speculative is completely unrealistic. There are a few situations where the advantages seem to be particularly clearly defined, and where a modified and limited version might be attempted. One such case is in central Mexico, where maize, throughout its history of cultivation, has grown in association with a weedy relative called teosinte, with which it hybridizes at a low but significant rate.

This exchange of genes has been crucially important to the maintenance of the genetic diversity of maize, and can be maintained only under conditions similar to those in which it has always occurred. In 1987 the United Nations created the Sierra de Manantlán Biosphere Reserve to conserve teosinte and the practices associated with its traditional cultivation.

The possibilities of *ex situ* conservation were transformed during the twentieth century by the discovery that seeds of many species could be kept alive almost indefinitely in cold stores. Formerly, collections of seeds could be maintained only by repeated rejuvenation every few years to replace deteriorating stocks. Every time this is done the plants are exposed to selective pressures, to the risk of hybridization with neighbouring collections, and to the possibility of misidentifications and the exchange of identities between different collections. These ill effects are almost eliminated by increasing the storage life of seeds to the point where they have to be regenerated extremely infrequently. All small, dry seeds, and that means the great majority, remain alive for very many years provided they are kept dry and at low temperatures (p. 132), and refrigerated cold stores make this an entirely practical, economical and highly effective option. Seeds of all the major crop plants are sufficiently resilient to survive temporary failures of refrigeration equipment, and the main threats to collection survival are war, political changes leading to the loss of financial support, or extreme natural events.

Seed banks, based on long-term cold storage, have now been established by numerous national, and a few international, institutions, as a resource for plant breeders and others for the benefit of humankind. The number of collections in these institutions provides an eye-opening introduction to the extent of the problem, and a measure of what an internationally coordinated programme to conserve the genetic resources of plants on a global scale has to deliver. By 2006 the seed collections in the National Plant Germplasm System in the United States numbered more than four hundred and seventy-six thousand, comprising nearly twelve thousand different species. Many of these are in the cold stores first installed in 1958 in the seed bank at Fort Collins. The N. I. Vavilov Institute in St Petersburg holds more than forty-three thousand collections of peas, lentils, beans, clovers, trefoils and other members of the pea family alone.

An early survey by the European Plant Breeders' Association estimated that any attempt to conserve crop genetic resources in Europe would need to make provision for at least one hundred and fifty thousand collections. The International Rice Research Institute at Los Baños in the Philippines, with its responsibility for a single, though vitally important, crop holds more than ninety thousand collections of different species and cultivars of rice. The seed depository on Svalbard, known as the Global Seed Vault (39, 40), is intended to hold some two million collections, which is deemed to account for the sum total of the world's plant genetic resources.

The first attempt by the Food and Agriculture Organization of the United Nations (FAO) to put the problems, possibilities and responsibilities involved onto an international platform took place in 1961 at a conference on plant exploration and exploitation at their headquarters in Rome. At the time there was more interest in making effective use of available global resources than their conservation. Not surprisingly, such recommendations as there were relating to conservation made little impact. In the absence of single-minded individuals prepared to negotiate the political hurdles and push through the smothering bureaucratic processes of a large international organization, recommendations were unlikely to make much impact. That situation was shortly to change.

Three years later an independent group of scientists set up the International Biological Programme to coordinate research on biological resources, the effects of environmental change and their impact on human societies. No organization with those aims could fail to take an interest in the dire consequences that changes in agriculture and other developments were having on the wellbeing of plants and animals. Scientists with relevant experience in many parts of the world were therefore enlisted to play a part in the programme's activities. One persuaded to play a part was Otto Frankel, head of the Department for Scientific and Industrial Research in Canberra, Australia. Frankel was a man who needed no convincing that greater efforts to collect and safeguard the rapidly diminishing genetic diversity upon which plant breeders depended were essential. During the next few years he would be a major player in the rapidly broadening interest in the problems involved, culminating, in 1967, in the organization of another international conference at FAO headquarters, this time in partnership with the International Biological Programme. The main theme of the 1967 conference was the exploration, utilization and conservation of plant genetic resources. The tone of the conference was set with the declaration that: 'It is deemed a national and international obligation to discover, conserve and make available the world's plant genetic resources to all who at local, national or international level may profit man by their access to them.'

Frankel found influential allies in Monkombu Sambasivan Swaminathan, Jack Harlan and, crucially within the walls of FAO's Rome headquarters, in Erna Bennett, who worked in the Crop Ecology and Genetic Resources Branch. I first met Erna and Otto, the name he instructed everyone to call him by, a few years later, having what appeared to be a flaming row in a lift in the FAO building. Erna was ticking Otto off for pressing on with one of his plans without consulting her, remarking that if he really thought he could rush in and persuade officials at FAO to accept it he was even more foolish and out of touch than she had thought. Frankel countered by telling her she was a stubborn, self-opinionated woman who was impossible to work with, and it was time she listened to what others had to say for once. This exchange, just one among many animated discussions between two people with strong opinions, encapsulated the strong and weak points of the characters of both these remarkable people. The combination of one with the other was largely responsible for their success. During the next few years both would play leading roles in establishing the conservation of the genetic resources of crop plants as a priority for coordinated international action.

Otto Frankel, born in Vienna in 1900, was described in a biographical note by Lloyd Evans, president of the Australian Academy of Sciences, as a man with 'a complex personality that could be rough or kindly, bored or engaged, impossible or altogether charming by turns'. Such a mercurial temperament, combined with an outstanding intellect and a constantly enquiring, disputatious attitude to everything he did, ensured that he would always be a maverick among his peers.

After exposure to various more or less unfulfilling educational experiences, the course of Otto's future was decided during a lecture on plant genetics by Erwin Baur, director of the Kaiser Wilhelm Institute for Breeding

Research in Berlin. Otto, still a student, was fascinated by what he heard. Baur found him a research project and Otto was launched into a career as a geneticist. His first job, breeding improved varieties of wheat and barley on a large estate near Bratislava, provided a framework for continued studies on plant chromosomes. This combination of the practical and the academic, which ideally suited Otto's temperament and capabilities, would be the hallmark of all his future activities.

After working for a time on oats and wheat at the Plant Breeding Institute in Cambridge, he was offered a post as plant breeder and geneticist at the Wheat Research Institute at Lincoln in New Zealand. During the next twenty-two years Otto Frankel produced new varieties of wheat adapted to conditions in New Zealand. On a visit to Britain in that period, he found an opportunity to work with Cyril Darlington, the renowned geneticist, at the John Innes Research Institute. He also travelled to Leningrad in 1935 to spend a week in the company of Nikolai Vavilov. Like so many others, Otto fell under Vavilov's spell and was converted to his approach to the collection and conservation of crop plants and the utilization of their genetic resources. The experience would later have highly significant consequences.

Frankel left New Zealand in 1951, suffering from a cultural famine (expressed as a lack of 'old stones and modern art') and frustrated by administrative bureaucracy after promotion to director of the Agronomy Division of the Department of Scientific and Industrial Research. His next challenge would be in Australia, where he became chief of the Commonwealth Scientific and Industrial Research Organisation (CSIRO) in Canberra. His first impressions were unfavourable; buildings were run-down and staff were demoralized and lacked leadership. During the next few years Frankel presided over a considerable shake-up, rejuvenating both buildings and staff, bringing new blood into the organization, inspiring long established members and introducing new departments and initiatives.

In 1966 he returned to the laboratory as an Honorary Research Fellow, but retirement was not an Otto concept. For the next thirty years, almost until his death at the age of ninety-eight, he would be directly involved as a leader and initiator in the movement to achieve international recognition of the dire threat to the genetic resources on which the plant breeders of the world depended. A lifetime's involvement in plant breeding, augmented by experience of top-level administration of scientific research, made him formidably qualified to play the part of gadfly. However, the flexibility of his approaches, his tendency to add new idea to new idea and the scope of his imagination out-paced the laborious processes by which things are decided in large organizations – inevitably leaving the bureaucrats with whom he dealt out-paced, baffled and frustrated.

Erna Bennett was the partner Frankel needed to enlist the support of FAO officials. Born in Northern Ireland, Erna was never less than fully committed to anything she did. At the FAO she initiated an international survey of crop germplasm as part of her responsibilities for coordinating collections of seeds, landraces and obsolete varieties of crop plants around the Mediterranean, and across the Middle East to central Asia and Afghanistan. In the course of her work, Erna coined the phrase 'genetic conservation', and is credited with having brought substance and strategy to the campaign to preserve genetic resources. Her job gave Bennett immediate access to, and familiarity with, those in the organization who made decisions, and she proved much better than Otto at finding her way through the snakes and ladders of the bureaucratic games by which decisions are made.

The 1967 conference moved conservation to centre stage. A review of the situation concluded that genetic erosion was already well advanced, and the consequences 'may gravely affect future generations, which will, rightly, blame ours for lack of responsibility and insight'. The recommendations of the conference set out an ambitious programme charging FAO with responsibility for coordinating national and international efforts to explore, conserve and make use of plant resources. This good intention was given teeth through a panel of experts made up of scientists familiar with particular crops or geographical regions. It was their job to keep abreast of current developments and push things forward by setting up working groups responsible for particular geographical regions, crops or techniques.

During the final years of the 1960s into the early 1970s progress was being made, despite Otto's impatience with what he felt was bureaucratic inertia on the part of the FAO. Preliminary surveys revealed the extent of the losses of genetic resources and stressed the urgent need for action. The panel's response, at a meeting in Rome in 1970, was to propose the setting up of a global network of centres responsible for the long-term conservation, propagation and distribution of plant breeding material and new varieties. This was followed, during a meeting of a working group at the USDA's research centre at Beltsville in Maryland, by identifying institutions suitable for inclusion in the network. Most of the institutions already existed, and many held substantial collections of seeds or plants. The choice was strongly influenced by the need to conserve what was left in the 'centres of origin' identified by Nikolai Vavilov and Jack Harlan.

A third conference was organized by FAO in 1973 'to review progress and define future objectives'. Otto was in the chair, determined to instil a little more urgency into FAO's activities and not least to redress the foot-dragging progress he attributed to his predecessor. I attended the meeting as the junior member of the British contingent, and found myself plucked from obscurity to act as the appointed leader – as the only civil servant present, I was deemed to be the only one with an official position! I was made chairman of the committee responsible for drafting the recommendations at the conclusion of the meeting. The recommendations that emerged were concerned mainly with putting into effect the panel of experts to establish a global network for the long-term conservation of crop genetic resources.

Other institutions, directly involved with plant conservation complementary to the FAO network, though not part of it, included the Japanese National Seed Storage Laboratory, CSIRO, Australia, and the Central Institute for Research on Cultivated Plants in East Germany. Other organizations and governments maintained networks to conserve what was left of the old landraces, redundant cultivars and breeding material. These included the N. I. Vavilov Institute of Plant Industry, which continued to coordinate activities throughout the Soviet Union, and the National Program for the Conservation of Crop Germplasm in the USA. The latter's facilities included the National Seed Storage Laboratory in Colorado, then, as now, the largest refrigerated seed bank in the world – though whether it will be overtaken by the Global Seed Vault remains to be seen.

In 1971 FAO gave birth to the International Plant Genetic Resources Institute, designed to coordinate activities within and between different seed banks and centres of research. This metamorphosed into Biodiversity International with headquarters in Rome and more than three hundred staff deployed around the world in some sixteen locations. In 1975 there had been fewer than ten institutions worldwide adequately equipped for the long-term conservation of genetic resources. By 1982 there were thirty-three, and twenty-five years later collections of one kind and another were being maintained in more than a thousand locations. In the course of a century, the movement had grown into the greatest coordinated enterprise concerned with conservation in the world.

The European Plant Breeders' Association set up an International Gene Bank Committee to coordinate genetic conservation in Europe and the Mediterranean region, especially the long-term conservation of seeds. Jack Hawkes, one of FAO's Panel of Experts, and professor of plant taxonomy at the University of Birmingham, was the chairman, and the Committee itself was made up of representatives from each of the institutions involved. Hawkes was renowned for his expertise in the improvement and conservation of potatoes, the only major temperate crop in which seeds play no part as far as most of those who grow them are concerned, although they are vitally important for the breeding of new varieties.

Hawkes had also been inspired by meeting Vavilov, and his memories of the encounter were to stay with him throughout his career. He had been involved in numerous collecting expeditions to South America in search of potatoes, one of which had lasted eight months, from late 1938 to mid-1939 – an unusual experience nowadays when short, sharp expeditions lasting weeks rather than months are more usual. He had also lived and worked in the Andes for three years, giving him time and opportunity to become fully familiar with patterns of growth and the geographical distribution of his quarry. In addition he had a most equable, friendly approach to those with whom he worked and this, allied to a great deal of common sense as well as academic distinction, made cooperation with him a pleasurable as well as an enlightening experience.

I owed my place on the Committee to the seed bank that I had set up at the Royal Botanic Gardens, Kew, before moving it to Kew's satellite garden at Wakehurst Place in Sussex. Seeds were a crucial element to the plants represented in the living collections at Kew and similar gardens worldwide, as European botanic gardens had been exchanging packets of seeds since the seventeenth century. Consequently many botanic gardens produced lists of the seeds that had had been harvested from the plants in their gardens; in the early 1960s Kew's annual seed list, *Index Seminum*, comprised thousands of entries. By the mid-1960s, however, it had become clear that many of the samples being distributed by Kew (and other botanic gardens worldwide) often failed to contain usable quantities of good quality, accurately identified seeds. Instead seeds were often wrongly named, dead or even non-existent.

Standards had to improve if the effort devoted to seed collection was to be rewarded. Yet seeds in Kew could be collected only between the middle of June and the end of October. Everything had to be done in a rush and under pressure in order to send seed lists out in time to be useful, leaving simply not enough hours in the day to check quality and weed out mistakes. To create the additional time required, it was necessary to store dry seed at low temperatures. By the end of the 1960s a fledgling seed bank, the Seed Unit, was functioning at Kew and, although the number of entries in Kew's *Index Seminum* dropped dramatically, their quality increased.

At Kew it was becoming clear that accessions to the Seed Unit, and to Kew Gardens more generally, should be based on the principle that the samples should be of known wild origin, and supported by detailed records of where they had been collected, by whom and when. They should also be accompanied by descriptions of neighbouring vegetation, the nature of the soil, the climate and other characteristic features of its habitat. The acquisition of the mansion at Wakehurst Place by Kew gave the Seed Unit the space to expand into a seed bank, and to take on broader responsibilities as a gene bank focused on the conservation of wild plant species.

The Seed Bank's original purpose, directed towards improving the quality of *Index Seminum* from seeds produced by plants in the collections at Kew, was history. The new challenge was to set up and run a seed bank devoted to the long-term conservation of the seeds of the world's natural flora. In the face of threats to the world's flora, seed banking provides the only practical and feasible way to conserve this diversity. It is not a

37 Redcurrant tomatoes have become popular fruits in their own right. However, wild redcurrant tomatoes have proved to be important sources of genes for resistance to the fungal infection fusarium which often infests commercial tomato plants.

38 Black stem rust is a serious fungal disease of cereals, and is a particular concern for wheat farmers in developing countries. The brown pustules of fungus covering a wheat plant shown here will eventually liberate more spores. Infected plants produce few grains and in severe cases will be killed.

39 Seeds, if kept cool and dry, may be stored for hundreds of years in seed banks. Here small seeds are sealed into glass vials for storage.

40 The Svalbard Global Seed Vault is located close to the North Pole. The unassuming entrance hides the seed vault which is carved about 120 metres (130 yards) into the side of a mountain, some 130 metres (140 yards) above current sea levels. The facility has the capacity to conserve 4.5 million seed samples.

41 Conservation of species in seed banks needs large amounts of high-quality seed. The seed collection from wild plants is a professional activity which requires thorough preparation before expeditions. For example, it is necessary to get permission to collect, and to understand the flowering, fruiting and distribution of the target species. Once in the field, seeds need to be correctly identified, collected and recorded before they are finally incorporated into the seed bank.

42 Kew's Millennium Seed Bank Partnership is the world's largest *ex situ* plant conservation programme. The Seed Bank achieved its goal of conserving 10 per cent of the world's plants by 2010. The challenge now is to store seed from 25 per cent of the world's wild plants by 2020.

43 Large seeds present a challenge for storage in seed banks since conserving many large seeds requires a lot of space. Here seeds in Kew's Millennium Seed Bank at Wakehurst Place are shown stored in familiar preserving jars at low temperature.

substitute for *in situ* conservation by preserving and maintaining the habitats in which the plants grow, but it does provide an effective and economical means of *ex situ* conservation.

Jack Hawkes had collected samples of the innumerable potatoes that grow along the Andes and the Pacific seaboard of South America, some in the fields of the farmers, others wild. His experience of using genes of the latter to introduce disease resistance into commercial stocks had made him more aware than many people of the value of wild species, and he played a very significant part in ensuring that the Seed Bank at Kew was represented on the Committee. Other members, whose experience revolved around cultivated plants, regarded the idea of banking the seeds of wildflowers with some surprise. They thought it eccentric to conserve wildflowers in this way, rather than plants with more obvious economic significance, and looked on Kew's efforts with a certain amount of good-humoured amusement.

One visitor to the seed bank after its move to Wakehurst Place was Lawrence Hills, founder of the Henry Doubleday Research Association at Ryton, near Coventry, now known as Garden Organic. He wanted to save stocks of long established vegetables in Britain that had fallen foul of European legislation and were fast becoming obsolete.

At one time, as anyone who set up extensive trials of vegetables soon discovered, the name of a variety on a packet of vegetable seeds was often only an approximation of what the seeds in it would produce. Half a dozen packets of cauliflower seeds – all bearing different names – might produce plants that came up identical in every respect. Alternatively, plants obtained from seeds supplied by different seedsmen of a particular variety of broad bean might bear little resemblance to each other. When Britain entered the European Common Market, an international organization to promote economic integration, in 1973, it eventually became subject to regulations designed to deal with these problems. A licence would be required before seeds of any vegetable could be sold. Licences were granted only when the inspectors who oversaw the regulations were convinced that seeds were accurately named, and would produce crops that matched carefully defined characteristics for the variety in question. Anyone caught selling unlicensed seed was liable to a very large fine. Synonyms and multiple identities disappeared, and no one, apart perhaps from a few seedsmen, mourned their passing, but the cull also had less desirable consequences. Licences were costly and had to be renewed periodically, making them expensive to maintain. It was only economic to license varieties which attracted large sales, and those which sold in small quantities began to disappear from the catalogues. Long established stand-bys of allotment holders and private gardeners faced the imminent prospect of being lost. Lawrence Hills's mission was to save these threatened varieties, and make them available to those who wanted to grow them.

Their salvation presented few problems, and Lawrence was persuaded that cold storage provided a safe, effective and economical way to keep the great majority of the seeds in which he was interested alive and in good condition for many years. Finding ways to make them available without running foul of European regulations needed more ingenuity. The vegetable seed bank at Garden Organic is a fine example of the use of a seed bank, on a modest scale, to achieve a particular and worthwhile result, and counterbalances any impression people may have that the technique is suitable only for grand scale projects to conserve resources of enormous economic value. It is entirely feasible and practical to conserve seeds for very many years by doing nothing more complicated than drying them over safe desiccants such as silica gel and storing them in airtight containers in cold stores. The main limitation of the method lies in the fact that only seeds that can be dried can be stored at sub-zero temperatures. Large moist seeds (nuts, acorns and the like) contain between 30 and 50 per cent water and die when they dry out; if not dried out, they freeze solid and perish at temperatures below freezing point. Fortunately, the great majority of the seeds that farmers and gardeners find interesting do withstand drying extremely well.

It is now well over a century since the secretary of the infant USDA, James Wilson, set up the Section of Seed and Plant Introduction under David Fairchild's leadership at Beltsville in Maryland. The Section's remit was to make collections of seed of useful plants, conserve them for future use and distribute them as breeding material. The ambitions and responsibilities of the establishment have expanded beyond imagining, and the use of refrigerated cold stores in the seed bank at Fort Collins has transformed its effectiveness, but its aims have scarcely changed at all. In recent years the USA's National Plant Germplasm Scheme has been distributing more than a hundred thousand samples of seed a year to research workers, plant breeders, pharmaceutical companies, gardeners and others in more than a hundred countries.

The conviction that genetic resources are the common heritage of mankind has been a fundamental clause in the USDA's creed, and one which for many years was accepted, either actively or passively, almost everywhere. Vavilov sought no permission to collect seeds from the governments of the countries he visited, and his relations with botanists and plant breeders in these countries were mutually supportive but informal. When Frank Meyer went on his walking tours throughout China and across much of central Asia, he was collecting seeds and other material from plants to enrich the agricultural resources of the United States. Although Meyer would have sought permission from landowners and peasant farmers to collect from their fields, and permits and passports from local magistrates and other authorities to travel through the country, he would not have thought of applying for official permission from governments to collect seeds and plants. Most of them would have considered him quite mad had he done so, and doubtless would have refused his application. Even in 1970, when Chris Humphries and I went to the Balkans on behalf of Kew's Seed Bank, there was no question of seeking formal approval to make collections from the governments of Yugoslavia, Greece and Bulgaria.

Today we would be condemned as biopirates for such irresponsible behaviour. The easy acceptance that plants were free for all, and that those who gathered and made use of them earned exclusive rights to benefit from them in whatever way they could, has been challenged. Although the great majority of the collections made never acquire much, if any, value (often proving a cost rather than an asset), a few do turn out to be enormously precious. The genes which conferred resistance to stripe rust on the unimpressive wheat plant collected by Jack Harlan in Turkey were later incorporated in virtually every variety of wheat grown in the northwestern United States. Some collections of grasses, clovers, trefoils and other forage crops have been used as they are, without the need for modification, to improve the quality of grazing over tens of thousands of hectares of pastures and rangelands. Many pharmaceutical products, first obtained from plants collected in the developing world, have made fortunes for companies in more developed economies. Seeds collected in Brazil by Henry Wickham (an adventurer with an eye for the main chance) and sent to the Royal Botanic Gardens at Kew in 1876 produced the plants from which the rubber industry in the Far East grew. Not surprisingly, many countries developed a feeling that they were being exploited, and the conviction grew that ways should be found to share the benefits more equitably.

Few would argue that this is no more than natural justice, but the issue has been extremely difficult to resolve satisfactorily. The USDA's creed, shared by many others, that genetic resources are the common heritage of mankind is something that works best among equals. It can seem no better than a recipe for exploitation when applied by rich nations with much to gain to poorer ones who feel they have something to lose. More formal arrangements were needed to balance what was seen as the gap between those who collected and those who provided, to ensure that the latter received their dues. Early attempts to redress the balance tended to trade permission to make collections against a share of whatever was obtained. Only a very few years after my visit to the Balkans, collectors from Kew, as elsewhere, were having to strike bargains of this kind. At the time of my Balkan trip it would have been a pointless waste to leave half of hard-earned collections behind in countries which had neither the trained staff nor the facilities needed to make use of them.

The situation was brought to a head when more than 180 countries signed the Convention on Biological Diversity following the 1992 United Nations Earth Summit in Rio de Janeiro. By doing so they recognized the principle that nations had sovereign rights over the plants and other natural products within their countries, and that such resources could be collected and exploited only with the consent of the governments concerned. The Convention stressed the need to recognize the value of such resources and to make arrangements that reflected this when reaching agreements. It also drew attention to the need to respect and conserve traditional knowledge concerning plant resources, as well as pointing out the critical, and often overlooked, role of women in the preservation and utilization of crops and wildflowers. The Convention was a significant landmark among international agreements. It invoked principles of ethics and equality for the first time, and the provisions were legally binding on the signatories.

The apparently glaring failure to recognize the contribution of peasant farmers and their forebears, whose efforts had maintained the landraces that had become so valuable to plant breeders, was one of the first issues to be addressed. A movement developed for the recognition of what were referred to as 'Farmers' Rights', based on the view that legislation conferring rights on plant breeders to benefit from the varieties they produced might provide a solution. The ways in which the legislation was framed frustrated attempts to do this and the whole question was conveniently set aside by declaring it to be a special case which required special provisions. Years of legal wrangling and political posturing over what those provisions might be – and how they might be applied – followed, during which the practical problems of allocating benefits in a fair and satisfactory way to such a nebulous group of people became more and more obvious. Prospects of progress became increasingly remote as protection for intellectual property rights became ever more restrictively defined – and ever more the province of large, well organized, often multinational organizations. Eventually the issue of Farmers' Rights became so marginalized, restricted and reduced by legal arguments, and beset by practical ones that it was virtually suppressed.

The failure to secure rights for peasant farmers contrasted with the growing realization of the value of plant genetic resources. The potential benefits to be gained from plant protection legislation were becoming evident to other, more organized and powerful vested interests, determined to take advantage of the opportunities offered. Coincidentally, with the extension of the Green Revolution to parts of the world where high-tech agriculture had never been seen before, legislation was being created to safeguard the interests of plant breeders and reward their efforts. This raised prospects of vast profits for those able to produce not only new varieties of wheat, barley, maize, rice and other crops, but also the ancillary products – fertilizers, herbicides and pesticides – that supported their growth.

Big multinational corporations became major players in a field that had previously been almost exclusively the preserve of governments. Their entry was greatly facilitated by a worldwide decline of public sector support for plant breeding, and an increasing inclination to contract support for research projects of all kinds to the private sector.

The issues raised by this development exerted pressures on the world-wide efforts to bank the genetic resources of plants and make them freely available to those who needed them. It threatened to subvert the whole philosophy of service for the public good on which they were based. In Rome, Erna Bennett became so concerned by the increasing success of multinational companies to exploit intellectual property rights for their own interests that she resigned from her job at FAO in 1982.

The latter part of the twentieth century has been the setting for a pervasive weakening of belief in the concept, exemplified by the collection and distribution of plant genetic resources, that knowledge is a common good and the free exchange of information between scientists a vital social obligation. The conviction that we are the custodians of nature and its wealth, but that it is not our property has been undermined by equally strong beliefs in the sanctity of property rights. Values based on the ethos of our role as custodians have been eroded by an increasingly uncritical acceptance of the rights of private and corporate interests to profit from whatever situation in almost any way that they can. The substitution of private profit for common good as the driving force behind an activity so fundamental, not just to human wellbeing but to human survival, has created an extremely hazardous situation, and one which could all too easily lead to disaster.

By the 1990s it became clear that if the Kew Seed Bank was seriously to meet its aspirations to conserve threatened plants across extensive swathes of the globe, it needed the staff, facilities and financing to match. At about the same time the Millennium Commission stated its intention to make substantial grants to projects deemed appropriate to mark the start of the third millennium. An establishment that claimed to be able to conserve plants to the dawn of the fourth millennium seemed to have a chance of funding. The outcome was an application for the funds needed to transform the Bank into a major international facility, operating on a budget, and on a scale, that would dwarf what had gone before. The target was to collect the seeds of some twenty-four thousand species of wild plants by 2010 (see p. 256). These would be obtained from many sources through partnership arrangements with botanical and other organizations in the countries where they grew. A very strong emphasis was laid on the importance of making equitable arrangements which ensured that a share of any seeds collected, as well as training and expertise in seed banking, plant conservation and the more effective use of plants, would be given to local participants. Seeds from many sources would be included in the operation, but with an emphasis on those from the world's drylands, where human welfare, more than in most places, depends on traditional uses of local plants. The project was awarded nearly thirty million pounds in December 1995, on condition that this was matched by complementary funding from elsewhere. The Millennium Seed Bank (42, 43) was officially opened on 20 November 2000.

Seed banks were set up to collect, conserve and distribute the material on which plant breeders and thousands of others engaged in the study of all aspects of plants depend. The pros and cons of the greater involvement of major business enterprises is officially of no concern to those who run them, though views on the subject, such as those held by Erna Bennett, may influence the actions of those who work for and with them. Nevertheless, seed banks are, or should be, concerned with ensuring that any benefits are shared equitably with the donor nations and, if possible, individuals who provided the collections. Indeed, the provisions of the Convention on Biodiversity legally bind almost all of them to do so.

Attempts to reward and protect peasant farmers for their part in producing the landraces that contain the reservoir of genes for past and future crop improvement foundered on the rocks of legal intransigence. But the benefits of seed technology, not least to those concerned with banking seed for future use, can be equitably distributed much more directly, particularly in the more undeveloped countries where plants of many kinds are used to nourish and support local people.

Nikolai Vavilov and David Fairchild regarded collections of seeds primarily as a means of making plant genetic resources readily available to the plant breeders of the Soviet Union and USA respectively, and their seed banks were run almost exclusively for the benefit of the more technically developed societies. Then, as we have seen, the possibilities of using seed banks to safeguard these resources, and to conserve them indefinitely for future use, led to a massive international endeavour. By 2006 this effort was credited with having safely gathered in and conserved 95 per cent of the available genetic diversity of rice, wheat and maize, along with substantial proportions of other major crop plants. Now in the third phase of banking, we are seeing increasing interest in gathering together and evaluating plants whose economic uses and value are still more or less untapped. This provides opportunities to apply the benefits of seed banking, and the application of more sophisticated forms of cultivating and exploiting plants, to people in developing countries. This will be an interesting enterprise, as plants are used in a multiplicity of ways, far removed from the norms of those accustomed to the economics of wheat, maize and rice.

Future Prospects

In 1798, in his influential book *Principles of Population*, the Reverend Thomas Malthus observed that 'the power of population is indefinitely greater than the power in the earth to produce subsistence for man'. In other words, he believed that the human population is capable of growing much faster than its ability to produce adequate food, and that it would outstrip the available food supply by the middle of the nineteenth century. This conviction, and the conclusions Malthus drew from it, provoked arguments that have continued to the present day. Charles Darwin and Alfred Russel Wallace, the great evolutionary biologists of the nineteenth century, saw within it the means by which variation could be maintained and eliminated from populations. However, the dire predictions that Malthus made for people and their food supply did not happen. Scientific agriculture intervened to ensure that food production kept pace with human population growth, assuming that food was equitably distributed.

There are more people on Earth today than ever before, all requiring plants for food, shelter, fuel, medicine and the other essentials of life. However, we are also faced with circumstances where major man-made environmental changes will happen within the next century, plants will be driven to extinction and the natural variation of crop plants and their close relatives will continue inexorably to be eroded. The question naturally arises whether our exploitation of plants will keep pace with our increasing population, and whether we have done enough. There are no definite answers. Yet this book has described the efforts that have been made to try and ensure that the evolutionary capital of the plants used by man will not be squandered.

For plants to adapt to changing conditions, they must inherit variation upon which selection can act. Furthermore, plants, and humans, need time to adapt to changes; otherwise, humans have to mitigate their own influences. Time is an important evolutionary commodity. The plants that we use today are the products of thousands of years of crossing and selection in different environments. In a completely different context, French historian Fernand Braudel thought of time as having three levels, explained through the metaphor of oceanic movements. The first level, the vast oceanic movements such as the Gulf Stream, represents the environment and is associated with slow, almost imperceptible change. The second level, tides and swells, represents cultural and social change. The third level, the spray on the surface of waves, represents rapid change, which in the human context is usually political. If Braudel's ideas are applied to the variation we see in plants, the first level is that associated with the natural distributions of wild plants, the opportunities for crops to evolve from their wild relatives and the ecological ranges of different plant groups. The second level comprises crop forms selected by different groups of people and the interactions of these forms with their wild relatives, the landraces. The third level encompasses chance mutations in natural populations and the brief breeding lives of modern crop cultivars. At each of these levels, seeds are the means by which genes are protected for transmission to the next generation.

A generation has passed since most of the events related in this book took place. From the 1980s technological and intellectual advances, indeed revolutions, have been breathtaking. Computing has become cheap and high-speed, while global communication systems, enabling huge quantities of information to flow between computers, are commonplace. Advances in computing have brought images of the entire planet to one's desktop at the push of a button, and global positioning devices enable us to know our precise location on the planet. Photographs are taken at virtually no cost, and air travel to most capital cities is quick and unsustainably cheap.

During the same period the evidence for Darwin's great idea – evolution – has become incontestable. We can unlock the DNA code from any organism on earth, and consequently our understanding of the tree of life has improved dramatically. We are now able to investigate the origins of species and the routes by which they have been distributed through time and space. Furthermore, we can rationally manipulate the genetics of different organisms and even create simple life forms from scratch. We have more information about the natural world than ever before, but we have only scratched the surface of what remains to be discovered. However, serpents remain within this intellectual paradise. The very environment upon which we rely, and from which we have striven to divorce ourselves over the last ten thousand years, is in crisis.

Gene banks, of which a seed bank is one type, are means by which plant variation may be conserved into the future. They are arks of variation, seeking to protect against the extinction of genetic diversity, and they are filled with plants – often many hundreds of different samples of each species. The Millennium Seed Bank, located in southern England, caught the public imagination in the late 1990s, and the Svalbard Global Seed Vault, located close to the North Pole and carved out of a Norwegian mountain, is the latest example of an international seed repository. Yet seed banks, if they are not properly managed, are in danger of becoming memorials to the conservation ideal: mere vaults of unrealized potential, museums of species that once existed. Local seed banks have been criticized because of their lack of scientific rigour, yet the scientific rigour of international collections may mean that it is impossible to conserve some seeds. Heritage seed banks, with their remit of looking back to the past, have ensured the survival of seed varieties that do not pass muster today, even though they may contain genes of future importance. Worse still, seed banks may become temples at which governments pay their devotions, confident that they have done enough to secure humanity's future.

The focus of this concluding chapter is on the major environmental challenges that humans will face during the next century. Prospects are bleak, but plants might help us mitigate, or adapt to, these changes. With a finite amount of land, four issues will compete to increasing degrees: land for living, land for food, land for fuel and land for conservation. Worldwide four crops (wheat, rice, maize and sugar) provide 60 per cent of people's calorie intake and the use of plants as direct sources of liquid fuels is under active investigation. New ways will need to be found to manipulate traditional crops, and new species will need to be investigated as crops. Yet these approaches have already provoked diverging opinions about plant genetic resources and genetic modification, and have raised concerns about the rights of the peasant farmers as custodians of plant variation and plant genes as intellectual property (p. 250).

Twelve thousand years ago, as wheat and barley were first being domesticated in the Near East, there were between one and ten million people on Earth. By the time of Christ the human population lay somewhere between 170 and 400 million, and at the start of the English agricultural revolution, in the late eighteenth century, it had risen to between 800 million and 1.1 billion. In 1859, when Darwin's book *On the Origin of Species* was published, the figure had risen to between 1.1 and 1.4 billion. By the time Vavilov died, in 1943, the world's population lay between 2.4 and 2.6 billion, and the estimated figure today is 6.9 billion. By 2050 it is expected that there will be 9.1 billion people on the planet.

However, the human population is not evenly distributed. Asia accounts for approximately 60 per cent of the world population with almost 4.1 billion people; China and India alone account for about 37 per cent of the world's population. The population of Africa is currently about 980 million people (around 12 per cent of the total) and is expected to rise to about 20 per cent by 2050. In contrast the European population is about 720 million people (about 10 per cent of the total), but is expected to fall to about 7 per cent by 2050. The North American population (around 5 per cent of the total) is expected to remain unchanged over the next forty years. The uneven distribution of the human population and its future growth is important because some of the most populated regions on Earth are areas of greatest plant species diversity.

In 1968 ecologist Paul Ehrlich re-rehearsed Malthus's arguments on population growth and predicted widespread famine in the 1970s and 1980s. Yet his predictions, like those of Malthus, were wrong. Both underestimated the impact of agricultural research, especially the Green Revolution (see chapter seven), on food production. Since the start of the Green Revolution in 1950, about 3.9 billion people have been added to the world's population, as against the 2.3 billion people added in the millennium before 1950. The dividend of the Green Revolution was a steady – and dramatic – rise in agricultural production: global grain yields increased by 250 per cent between 1950 and 1984. However, this rise was underwritten by breeding genetically uniform crops, and exploiting the products of prehistoric photosynthesis to manufacture fertilizers, pesticides and energy. The illusion of plenty and the ready availability of cheap food made some countries, including the UK, complacent about investing in agricultural research, which became increasingly the preserve of private enterprise. However, by the early years of the new millennium rising food prices started to change this view. The FAO has estimated that to keep pace with our currently growing population, global food production will have to increase by 75 per cent before 2050. Yet the rate of increase in productivity through improving agricultural practice is slowing, therefore new approaches and crop species will be needed to increase yields on the same amount of land.

Humans have transformed natural landscapes to accommodate food production. Forests have been razed to the ground, marshes and fenland drained, and savannahs and prairies ploughed. However, high intensity agriculture was not an idea of the 1950s; it had been practised for centuries. The sinuous terracing of mountainside in the Andes, China and southeast Asia are some of the great human engineering feats over the last two millennia (44). These community achievements were made by generations of farmers so that soil and nutrients were trapped and the crops could grow productively. Yet today wholesale habitat transformation has become a serious concern for many people.

The dustbowl, and subsequent migration of poor farmers, defined the USA in the 1930s (45). In the 1990s the consequences of Chinese hillsides, stripped of their forests, slipping into the Yangtze were devastating for tens of thousands of Chinese peasants. Indeed, destruction of the Amazonian rainforest has become shorthand for human mismanagement of the environment (46). Western artists and environmentalists often hark back to landscapes of wild places before man destroyed them. In his *Georgics* the poet Virgil portrayed the Italian landscape as perfect: 'Here spring is perpetual, and summer extends to months other than her own; twice a year the cows calve, twice a year the trees serve us fruit…no ravening tigers or savage brood of lion; no aconite deceives the wretch who picks it.' Yet the landscape Virgil described was largely man-made, and the influences of humans in the Mediterranean generally have been dramatic. Closer to home, the quintessentially English landscape evoked by Edward Elgar is the product of generations of farmers moulding the land. As with population growth, landscape transformation is not evenly distributed across the planet. The greatest amounts occur in developing counties; most developed countries have already decimated their natural habitats.

Currently about one-third of the planet's land area has been converted to agriculture, but much fertile agricultural land is also being lost. In the past forty years, approximately one-third of all agricultural land (i.e. one-ninth of the planet's land area) has been abandoned because of soil erosion. Fertilizers, derived from fossil fuels, provide a temporary fix to enhance the fertility of eroded soils, but topsoils cannot easily be replaced. Under agricultural conditions, it has been estimated that it takes about half a millennium to form about 25 mm of soil. Where fertilizers fail, agricultural land is simply replaced by the conversion of marginal and forest land – another short-term solution, as the productivity of soils under tropical forests has proved illusory. Yet the planet's wild places continue to disappear. More than 40 per cent of land area is currently being converted from natural vegetation; less than 25 per cent remains intact. The majority of land that is relatively unaffected by humans is in extreme environments such as the poles, the summits of high mountains and in deserts. By 2050 up to 90 per cent of the land area will be affected by human activities.

Maintaining human food supplies is directly dependent on soil fertility, clean water, energy and biodiversity. Yet converting land for agriculture to sustain an increasing human population places strains on water resources, energy supplies and the existence of other species. Forests stabilize hillsides, prevent soil erosion and act as water reservoirs. However, they reduce the amount of land available for agriculture, which consumes more fresh water than any other human activity. For example, an area of maize the size of an international football pitch needs about two and a half Olympic-size swimming pools of water to produce a harvestable crop. In the last decade the sustainable supply of energy has become a serious concern. Wood burning is the direct source of energy for much of the world's population, while liquid biofuels – such as ethanol and biodiesel – have attracted the strategic interests of Western governments because of concerns over climate change and threats to fossil fuel supplies. Growing fuels also has implications for habitat transformation, however, since either land is converted from food to fuel production or habitats are destroyed to grow fuel.

Justifications for habitat and species conservation are often couched in terms of the potential direct benefits to humans. Organisms other than crop plants are important for food production, and these too are affected as the human population grows. Natural enemies of crop pests may be eliminated and crop disease increase. Bacteria and fungi that fix nitrogen, break down dead plants or form soil may disappear. In 2005 German and French scientists estimated that the worldwide economic value of the services provide by insect pollinators, such as bees, was around €153 billion. As insect pollinators decline, so will food and seed production; humans cannot readily replace these services.

Man-made climate change is happening and its effects on the entire planet's life will be wide-ranging. The Earth's temperature is determined by the balance between heat gained from the sun and heat lost into space. Greenhouse gases (for example, carbon dioxide and methane) provide a barrier to heat loss and are crucial to all life on Earth, both plant and animal; without them the Earth would be considerably colder than today. However, the more greenhouse gases in the atmosphere, the warmer the surface of the planet will get, which in turn will change where different plants can grow. One of these gases, carbon dioxide, is taken up by plants and is locked away, through the process of photosynthesis, in a form that cannot affect the atmosphere. All land animals ultimately obtain their carbon by eating plants, so carbon is released back into the atmosphere when plants or animals decompose.

However, if organisms do not decay carbon is retained. Over about 350 million years, plants and marine organisms were crushed beneath ocean sediments to form fossil fuels such as coal, gas and oil. As the Industrial

Revolution gathered pace in the nineteenth century, large-scale consumption of fossil fuels started to take place. This carbon is released back into the atmosphere, increasing the overall concentration of greenhouse gases and consequently the surface temperature of the Earth is rising.

Carbon dioxide levels in the atmosphere can be measured, and are usually expressed as parts per million by volume (ppmv). Ice-cores, which trap small bubbles of air, are used to investigate what the earth's atmosphere was like in the past. These data show that before the end of the eighteenth century carbon dioxide levels were about 280 ppmv. By the late 1950s they had increased to 316 ppmv, and today they have reached about 387 ppmv. Over the last 650,000 years carbon dioxide levels have ranged between 180 and 300 ppmv, yet human activity has caused a similar range of variation in just one century. The effects on climate have been dramatic and well publicised. Over the last 150 years global temperatures have risen 0.76°C. The dozen warmest years within this period occurred between 1996 and 2009.

Predictions about future climate change are, of course, uncertain and based on the outcomes of climate models. Using the best information available, the Intergovernmental Panel on Climate Change (IPCC) expects mean global surface temperature to rise between 1°C and 6°C by 2100. Global warming will increase the severity of floods, droughts, heat waves and storms, while vulnerable coastal communities will be exposed to a sea level rise of up to 60 cm. These effects, together with an increased frequency and magnitude of extreme climate events, will reduce the security of people's water and food supplies. As the IPCC's more extreme scenarios are considered, the consequences of global warming become more severe. More than a billion people could be displaced by rising sea levels, while disruption of the North Atlantic Ocean circulation (including the Gulf Stream) could push Western Europe into cycles of very cold winters and very hot summers. Disease patterns of plants and animals may change. Billions of people are also likely to face extreme water shortages and hunger; economic consequences would be severe, leading to increased mass migration of humans and war. The effects of global warming will be felt most severely by the world's poorest people who, ironically, live in parts of the world with the greatest plant diversity.

44 The sinuous terracing of mountainsides in southeast Asia for rice production are engineering achievements made by generations of farmers so that soil and nutrients are trapped and the crops can grow productively.

45 *Above* The dust storms of the American Midwest in the 1930s were caused by severe drought coupled with decades of intensive farming without crop rotation and techniques to prevent soil erosion. Deep ploughing of the topsoil of the Great Plains killed the grasses that kept the soil in place, even during periods of drought and high winds.

46 *Left* Destruction of tropical forest is used as shorthand for the loss of habitats across the globe. Here the contrast between forest and the aftermath of clearing the trees is clearly seen. Tracks mark the cleared area and with the arrival of rains the shallow upper layers of the soil will be washed away unless plants start to recolonize.

47 *Opposite* This detail from *The Harvesters* (1565) by Pieter Bruegel the Elder shows wheat plants being harvested which are much taller than most wheat grown in Europe today. Modern wheat plants have large, heavy ears so must be short to prevent them from falling over. Furthermore, short plants mean that plant resources go into grain rather than straw production.

48 Over the past two decades, the focus of seed banks has shifted from cultivated species and their wild relatives for plant-breeding purposes to the conservation of wild plant species. Germplasm collections show the diversity of fruit and seeds produced by flowering plants.

Global warming will exacerbate species' extinctions, either directly or indirectly, since people will move to environments in ever more marginal habitats. Such environments have traditionally been given over to habitat conservation because humans have few other uses for them.

The complex issues that surround human population growth, habitat transformation, climate change and global warming have raised considerable controversy. Societies are likely to respond in two ways. They will either adapt or attempt to mitigate the effects, with the hope that the rate and magnitude of any adverse changes will decrease. Adaptation and mitigation depend on economic and environmental circumstances, together with the availability of suitable information and technology. However, outcomes are uncertain. The costs of mitigation are high, while the costs of adaptation are largely unknown. Uncertainty cannot be used as an excuse for inaction, and we cannot afford to be paralysed by complexity and doubt. If the costs of action are great, the costs of inaction will be greater still.

Opportunities for using biodiversity to mitigate climate change will be reduced as biodiversity losses increase through global warming. Furthermore, climate change will potentially exterminate the agricultural biodiversity that may help farmers cope with changing environments. The loss of the wild relatives of crops, necessary sources of genes for traits such as drought tolerance and pest resistance, is of particular concern. A study based on computer models predicted that by 2055 climate change would lead to the extermination of nearly a quarter of the populations of wild cowpeas, peanuts and potatoes, and that the remaining populations would be confined to small areas. As plant populations get smaller, the amount of genetic variation they contain is reduced. There is also a greater chance in small populations that plants will mate with close relatives. This will increase the number of deleterious mutations exposed to selection, leading in turn to further reductions in population size. The species will enter a spiral of decline (known as 'an extinction vortex') and eventually become extinct.

Much plant conservation, especially of economically important plants, is justified in trying to maximize the genetic variation preserved in a species, and to minimize the risk of the species entering such a vortex. Consequently much effort has gone into investigating the amount of genetic variation

241

captured in seed banks. Other activities include the estimation of gene flow within and among natural populations of a species and the monitoring of genetic mixing between wild and cultivated species, both of which are standard research practices.

In the 1920s the Soviet geneticist Nikolai Vavilov saw the importance of genetics for plant conservation and breeding, albeit through a glass darkly. The consequences of reducing crop diversity are dramatically revealed when new diseases evolve. In 1970, for example, an epidemic of southern corn blight resulted in the loss of 15 per cent of maize production in the United States. The epidemic started in Florida and moved north, but only affected maize containing a particular set of male sterility genes. However, since these genes had been used widely in maize breeding, the disease affected most of the US Corn Belt. To prevent the spread of such diseases, it is necessary to ensure that monocultures of the same genetic type are not created. Farmers in China improve resistance to rice blast fungus by growing several different rice varieties in the same field. The diversity of crop cultivars and their wild relatives are like an investment, put by for a rainy day; in times of difficulty, such as the arrival of a new disease, a breeder can dip into this genetic 'piggy bank'. The importance of such a genetic 'piggy bank' is illustrated by the fact that more than half of the new peanut varieties released to farmers in the twenty-first century have directly incorporated traits from wild relatives.

Genetic investigations have become powerful ways to study the natural world over the last two decades, particularly when combined with ecological analyses. Genetics allows us to explore in more detail particular aspects of the biology of the very limited range of traditional crops upon which human civilizations and lifestyles rely. As environments change, traditional crops will be required to adapt and new commercial crops will need to be sought. The hope of genetics is that the slow domestication processes associated with traditional crops can be bypassed in favour of either rapid re-creation and domestication of old crops or rapid domestication of new crops.

Discovering plant origins is more than an intellectual exercise. For wild plants, it provides opportunities to understand relationships between species. In the case of domesticated species, there is the opportunity to resynthesize a crop from its component species through a conscious, rather than unconscious, selection of particular traits known to be important in domestication. Rice is one of the important food cereals, and one of the crops for which it is believed that the majority of the current genetic variation has been captured in seed banks. The rice seed collection at the International Rice Research Institute contains over one hundred thousand accessions. Rice domestication is a process, but selection for a particular non-shattering gene resulted in rice becoming dependent on humans for survival, and vice versa. Identifying the number of Asian rice origins and the relationships of domesticated and wild forms has been the subject of much speculation, since few traits consistently differentiate wild and domesticated rice. However, recent genetic research supports a single origin of domesticated Asian rice in the region of the Yangtze river. As rice spread from its centre of origin, cycles of hybridization, selection and differentiation led to considerable genetic complexity. We rely on this inherited genetic complexity for breeding new rice cultivars today, and will do so into the future.

The selection of desirable genes or the elimination of deleterious ones is important for the development of improved crop varieties. Once the genes responsible for certain traits have been identified, breeding aids can be developed. Genetics can be used to speed identification of plants with desirable traits using a process called marker-assisted selection. DNA regions closely associated with the desired traits are used to select indirectly for the trait. This can be conceived by imagining a very large grain store containing a very small, steel needle inside a particular seed. You want to identify that seed. The quickest way of identifying the seed would be to pass the contents of the grain store under a strong magnet. DNA markers are like the steel needle and can be identified quickly using a suitable technique (the magnet). In real plant populations, marker-assisted selection is particularly valuable when the trait of interest is difficult or expensive to detect. Examples of this include leaf rust resistance in wheat, soybean rust resistance and drought adaptation in maize.

Traditional approaches to plant breeding involve the manipulation of the frequencies of particular genes for useful crop traits through processes of hybridization, and unconscious and conscious selection. This strategy has enabled humans to feed themselves for the last ten thousand years. However, the current challenges posed by climate change, increasing human population and habitat loss have led breeders to consider approaches that involve moving particular genes into crops or the re-creation of crop plants from their component species. These methods complement traditional breeding approaches.

In 1565 Pieter Bruegel the Elder painted *The Harvesters* (47). It is a remarkable work for many reasons, not least because of the height of the wheat plants that the peasants are shown harvesting. Wheat today is about crotch-height, in contrast to Bruegel's shoulder-height plants. The shorter wheat varieties produce better yields because their energy goes into filling grains rather than making straw; they are also less likely to fall over, further increasing overall wheat yield. Dwarf wheat is one of the triumphs of modern plant breeding, and was achieved by transferring 'reduced height' (Rht) genes from Japanese to western wheats (p. 197). Short wheat varieties originated in Korea as landraces in the third and fourth centuries AD, and were transported to Japan in the sixteenth century during the Korean–Japanese War.

Over the past two decades analyses of wheat DNA have shown that Rht genes have been spread around the world by three routes. The most important route for Rht gene dispersal was through the Japanese wheat variety 'Norin 10'. This originally arrived in the USA as a small sample of seed given to Samuel Cecil Salmon, a USDA wheat breeder, following a visit to a Japanese Research Station in 1946. In 1952 Orville Vogel, a USDA agronomist, crossed 'Norin 10' with 'Brevor', a popular wheat variety in the USA, to produce the variety called 'Gaines'. 'Norin 10' and its derivatives were transferred to CIMMYT, where new dwarf wheat varieties, insensitive to daylight, were developed by the Nobel Peace Prize winner Norman Bourlag in the late 1960s. These varieties were distributed globally, and increased wheat production in places as far apart as Mexico, India, Pakistan and Turkey: the Green Revolution was at full throttle. Two minor routes for the global spread of Rht genes were both from Japan via Italy. One was at the beginning of the twentieth century, with the transfer of Rht genes to south and central European semi-dwarf winter wheats in the 1950s. The other was from Italy to Argentina before World War II, and from Argentina to the rest of Europe and then the Soviet Union after World War II. Dwarfing would be advantageous to many crops, but wheat does not interbreed with many species, so Rht genes cannot be transferred. Using biotechnology techniques, dwarfing genes have the potential to be transferred to many other crops.

The term biotechnology used to describe processes that involve living organisms, such as cheese, bread and wine making, but today the word has become synonymous with genetic modification (GM). Public debate about GM technology in the context of agriculture became polarized, especially in Europe, in the late 1990s. On one side, the proponents argued that GM technology was no different to traditional plant breeding and was the only way to secure the future of human food production. The opposing side argued it was unnatural, that food produced by GM crops would be dangerous and the environmental impacts would be severe; the GM genie could never be returned to its bottle if it were once released. In the acrimonious debate of the early twenty-first century, one side sympathetically portrayed traditional breeders as mice spending their lives understanding crops in the field (these were the same breeders who were criticized for the Green Revolution of the 1960s). In contrast the new breeders were portrayed as rats confined to laboratories, with little understanding of crop plants. Just as 'organic' has become an unthinking shorthand for 'wholesome' and 'chemical' equivalent to 'toxic', so GM has also become associated with all manner of monsters in the public imagination.

The first GM plant was reported in 1983, but it was not until 1994 that the FlavrSavr™ tomato was commercialized. This tomato was engineered for longer shelf life, so that fruits could be kept on the vine to ripen and develop a flavour before being shipped to consumers. Tomato paste made from FlavrSavr™ tomatoes was sold in the UK, but withdrawn following the furore created by the naïve, inept marketing of GM maize imports into Europe at the end of the 1990s. In 2006 there were more than 100 million hectares of GM crops being grown by 10.3 million farmers in twenty-two countries. The main crops grown are oilseed rape, maize, cotton, soybean and alfalfa, and most of these are in the USA; almost none are grown in Europe. Apart from the limited range of GM crops, the number of commercial traits is largely restricted to herbicide tolerance and pest resistance. In 2006 herbicide-tolerant soybeans accounted for 87 per cent of the USA's soybean acreage. Sixty per cent of the cotton acreage in the USA was GM herbicide-tolerant and 52 per cent was GM pest-resistant. Thirty-five per cent of the total maize acreage in the USA was pest-resistant. The majority of herbicide-tolerant crops are tolerant to Monsanto's herbicide RoundUp™. Pest-resistant crops take advantage of genes from the soil bacterium *Bacillus thuringiensis*, which produces insecticidal proteins called Bt toxins. Different *B. thuringiensis* strains produce different Bt toxins, each of which targets particular insect larvae; some kill larvae of moths and butterflies, others beetles and bugs. Bt toxins are not toxic until they are broken apart in an insect's gut, where there are specific receptors that bind the toxins. This means that Bt toxins should be non-toxic to mammals.

Despite the limited numbers of genes introduced into commercial crops, many genes have been investigated in the laboratory. These genes are associated with traits as diverse as pest resistance, agronomic performance, stress tolerance and even the production of medicines. Pest-resistance traits improve crop performance by protecting the plant from fungi and insects. For example, a gene from a wild Mexican potato engineered into cultivated potato enables the GM potato to survive exposure to many races of potato blight, the fungus responsible for the Irish potato famine in the nineteenth century. Traits that improve field performance include crop yield and nitrogen-use efficiency, such as the gene introduced into the model plant, thale cress, that increased nitrogen content and improved growth under nitrogen-limited conditions. Stress tolerance, for example, high salt, high and low water availabilities, and temperature extremes, has been another target for genetic engineers. Some GM tomato plants have been shown to grow on approximately 40 per cent seawater, yet the tomato fruits contain very low concentrations of sodium. Changing the nutritional quality of plants has been most dramatically demonstrated by the creation of Golden Rice, which is enriched for vitamin A.

Globally vitamin A deficiency has enormous public health consequences. Between a quarter and half a million children with vitamin A deficiency become blind each year, and half of them will die within twelve months. Strategies for alleviating vitamin A deficiency have been developed by the FAO, including distribution of vitamin A pills in Nepal or sugar fortification in Guatemala, with varying degrees of success. However, incorporation of vitamin A into a staple food, such as rice, might be more effective. The first GM variety of Golden Rice had high levels of a vitamin A precursor, which was achieved by introducing three new genes, two from daffodil and one from a bacterium, into rice. A second Golden Rice variety used a maize gene instead of the daffodil genes. Golden Rice is not the complete solution to vitamin A deficiency worldwide, but both Golden Rice varieties are now being used in Asian rice breeding programmes.

Breeding new varieties of old crops using GM technologies requires enormous investment, in the same way that pharmaceutical companies must make huge investments to bring a new drug to market. Agricultural companies have therefore argued that they must recoup their costs, make a profit and protect their intellectual property. Methods known as genetic use restriction technologies have been invented to protect seeds of GM crop varieties. So-called 'terminator technology', a specific type of genetic use restriction technology, has attracted particular public and ethical attention since it was first patented in 1998. 'Terminator technology' involves inserting particular genes into a plant variety that interfere with protein synthesis and prevent germination of second-generation seed. Since fertile seeds cannot be saved from one generation to the next, farmers must purchase new seeds from the company each year.

Commercial farmers in developing countries, who grow conventionally bred, hybrid seed, are familiar with this aspect of modern agriculture. However, for farmers in developing countries who traditionally save seed and plant it the following year, growing crop varieties containing 'terminator' genes would mean that they could no longer practise traditional agriculture. Furthermore, these farmers would be tied to multinational seed companies. Other concerns include the potential that 'terminator' plants could cross-pollinate with non-GM plants and pass the 'terminator' genes into populations of closely related species. Following a storm of protest from lobbies associated with peasant farmers, indigenous peoples, conservation organizations and some governments, seed companies agreed not to commercialize 'terminator technology', although the technology still exists.

However, despite the potential social and environmental disadvantages, some have argued that the adoption of 'terminator technology' would stimulate companies to invest in breeding novel crops since they can protect their intellectual property more effectively. Others have argued that because seeds containing 'terminator technology' will be sterile, escape of genes into wild relatives, or the movement of genes from non-food crops into food crops, will be prevented.

The processes that have generated landraces, the foundations of future classical breeding, marker aided selection and genetic engineering efforts are the same processes that are the cause of such environmental concern when GM crops are discussed. Many crops are capable of crossing with wild or weedy relatives if their habitats and flowering seasons overlap. Crossing may produce new gene combinations that improve, harm or have no effects on the next generation of seed; this process makes landraces important. For crossing to be successful, parental plants must flower at the same time and be close enough together for pollen to be transferred to receptive stigmas and for viable seed to be produced, in due course to germinate and establish a new plant in the population. If the crosses are with GM herbicide-resistant crops, herbicide-resistant weeds may be formed. In Canada, crossing between herbicide-tolerant oilseed rape and weedy, herbicide-intolerant field mustard produces complex hybrid populations, where herbicide resistance genes have persisted in field mustard for over five years. However, herbicide-resistant weeds are not a concern unique to the intensive cultivation of GM plants. They may arise in any crop in any agricultural landscape. In conventional crops, herbicide-tolerant weeds are selected by incorrect use of herbicides or the transfer of herbicide-tolerance traits through outcrossing.

In the USA a study of pollen movement among GM herbicide-tolerant creeping bent – a wind-pollinated, outcrossing grass – and its wild relatives was conducted. It was found that during one season, the majority of pollen moved less than 2 km (1¼ miles) in the direction of the prevailing wind, with very limited gene flow up to 21 km (13 miles). Three years after the GM bent had been removed, 62 per cent of bent plants still contained the herbicide-tolerance gene. In this particular case, the herbicide-tolerance gene persists in the wild plants. Such observations show that genes can move between compatible plants, and GM traits can persist. This means that in areas where there is a high diversity of crop relatives, great caution is needed before GM plants are released. However, generalizations about the environmental or economic risks of gene movement cannot be made for GM crops. Case-by-case investigations are needed that focus on the crop, where and how it is to be grown and the trait of interest, just as they must for non-GM crops.

When the Mennonites emigrated from Ukraine to the USA in the 1870s, they brought with them both their culture and the seeds upon which they depended. Some of these seeds were from a particular hard winter wheat landrace known as 'Red Turkey'. Winter and spring wheats had been grown in the Midwestern United States since the 1830s, but they were vulnerable to drought, wind and dust storms, and cold, as well as fungal and insect pests. In the severe winter of 1896/97, 'Red Turkey' was one of only three wheat varieties to survive, and formal research subsequently revealed its high cold tolerance. Initially farmers did not want to grow 'Red Turkey' since it was unfamiliar and required changes to milling technology. However, a century later, two USDA agronomists confidently stated, 'This variety ["Red Turkey"] more than any other established the hard red winter wheat industry, answered for all time the critics who doubted the future of wheat as a crop on the Plains, and was the standard of quality on which the grain and milling industry of the Southwest was based. Turkey…has contributed an ancestral stamp to modern varieties, for no variety of hard wheat…lacks this lineage.'

Before the late twentieth century, legal and modern ethical concerns rarely inconvenienced plant collectors and breeders. Plants were the common property of humanity and powerful finders would keep all. Permits to travel safely, in sometimes hostile lands, were necessary, but once a seed collector was in the field any plants were fair game. Accusations of biopiracy have become the modern mantra of some environmentalists, politicians and lobby groups. Monochrome caricatures of seed collectors as the agents of 'ecological imperialism' have been augmented by parodies of plant collectors as Hercules stealing golden apples from the garden of the Hesperides.

Looking at Vavilov's map of the centres of plant diversity, it is clear that humans have always moved plants as they have migrated – and this continues to the present day.

The familiar garden strawberry first appeared in Brittany in 1740, as a cross between the tasty Virginian strawberry and the large Chiloe strawberry (p. 51), and rapidly replaced the tiny wood strawberry. After a hesitant start it is the garden hybrid with North and South American parents that is grown worldwide. European imperial powers proved particularly effective at acquiring economically important plants. For example, Britain tried to get living tea plants from China during the eighteenth century, and succeeded in the nineteenth century. In the late nineteenth century the Englishman Charles Ledger illegally collected quinine tree seeds from Bolivia and sold them to the British and the Dutch, laying the foundation of world quinine production until 1939. Ledger's collection was illegal since there were specific laws in Bolivia that prohibited the export of quinine plants and seeds except as part of a government monopoly. In the twentieth century Kew's reputation suffered through its involvement with the acquisition of rubber tree seeds and the destruction of the Brazilian rubber economy in the late nineteenth century. These attitudes were founded on the view that plants were the property of all people; whoever exploited them most effectively would reap the greatest rewards. The impacts of past biopiracy should not be underestimated just because the botanical rewards and the economic returns were so high. The social costs, in terms of poverty, warfare, slavery and exploitation, have often been enormous.

During the course of the twentieth century, there was a gradual change of attitude. The importance of sharing the rewards of plant discovery and the ethics of benefit sharing became recognized. However, it was not until 1992 that these attitudes crystallized in the provisions of the Convention on Biological Diversity (CBD) (see chapter eight). Finally most countries accepted that nation states had sovereign rights over the plants in their territories. Yet what this meant in practice has produced furious disagreements. Developing countries naturally wanted to share in the profits of commercially exploiting plant diversity, since much of the diversity came from their countries. In contrast, organizations exploiting plant resources wanted free access to the resources and opportunities to patent novelties. Complex chains of national and international ownership claims mean that identification of the beneficiaries of particular plants is often difficult to determine. The Mennonites had brought 'Red Turkey' seed to the USA from Ukraine, but they appear to have originally collected it from a region of Turkey. Yet no direct financial benefits have ever accrued to the Ukraine or Turkey for the genes that ensured wheat could be grown in the Midwest of the United States. Crops are the most dramatic examples of plants that generate considerable cash, the ownership of which may be difficult to determine. However, almost all the plants we grow in our gardens, houses and offices are the products of similar processes.

Applications for patents in developed countries – to plants and plant products that have been widely used in developing countries – are controversial. In 1995 the European Patent Office (EPO) granted the USDA and a multinational company a patent for an anti-fungal product derived from the neem tree. This was challenged by the Indian government, who claimed that neem had been used in India, as a fungicide, for over two millennia. In 2000 the EPO ruled in India's favour, but it was not until 2005, following an appeal, that the EPO revoked the patent. Another long running lawsuit, the case of the Mexican yellow beans, took fifteen years to settle. In 1994 a Coloradan seed company owner, Larry Proctor, bought a bag of bean seeds in Mexico and returned to the USA. Here he separated the yellow beans, planted them and produced more yellow bean seeds. In 1996 Proctor won a US patent on any yellow-coloured bean and obtained a plant variety protection certificate for a yellow bean he called 'Enola'. Armed with documents of legal ownership, Proctor started to sue companies selling yellow beans in the USA and blocking imports of yellow beans from Mexico. In December 2000, however, the International Center for Tropical Agriculture, home of the international bean seed collection, formally requested that the US patent be re-examined, and DNA evidence showed that 'Enola' was identical to other Mexican yellow-seeded beans. The case dragged on slowly through appeal after appeal until, in July 2009, the United States Court of Appeals ruled that the patent was invalid. Campaign groups were delighted, yet annoyed at the time taken to reach a decision in a biopiracy case that had such negative impacts on Mexican farmers.

Gene banks have been in particularly awkward positions regarding accusations of biopiracy. In 2001, under the provisions of the CBD, the International Treaty on Plant Genetic Resources was approved; a year later it had seventy-seven countries and the European Union as signatories. The Treaty recognizes the enormous contributions that farmers have made to crop diversity, establishes a global system that enables farmers, breeders and scientists to access plant genetic resources and establishes benefit-sharing protocols between the users of genetic material and the countries from which the materials originated. The Treaty has become the primary instrument guiding the activities of gene banks – although it has been criticized by some organizations for still not giving enough protection to the rights of developing countries.

Discussions of plant ownership, and the indigenous knowledge about these plants, may appear arcane. However, the potential rewards from having the right plant or plant product are enormous. The global markets of medicines, such as the childhood leukaemia drugs vinblastine and vincristine, both isolated from the Madagascar periwinkle, and taxol, an anticancer drug isolated from Pacific Yew, are worth hundreds of millions of pounds per year. Conservation must also be funded, and direct economic return arguments are often viewed as being particularly persuasive. The short lives of humans means that we are more likely to conserve plants with direct economic benefits rather than those with more abstract, long-term values: the 'use it or lose it' argument. Such attitudes may change, however, as the consequences of climate change start to take hold during the twenty-first century. The development of GM plants, and the prospect of moving genes across vast evolutionary distances, means that the natural world has become a genetic toy box of enormous economic potential. However, once a species has been recognized as having useful features the costs of exploiting it are vast. Sophisticated laboratories are needed, staff must be highly trained, plants must be bred and the products evaluated and marketed. Consequently most samples in gene banks will never produce a financial return – yet if a species' genetic resources are not conserved, they have a very high chance of being lost forever.

Conservation involves making decisions and using limited resources effectively. As a result conservationists are forced to put relative values on particular species. With their meagre budgets, how do they determine the relative importance of conserving weedy Middle Eastern grasses, southeast Asian tropical trees, Andean herbs, old cultivars of European vegetables and British dandelions? The problem becomes even greater when plants are in competition for conservation funding with charismatic animals. In 2004 a seed collection made in 1963, and stored in a seed bank, was used to re-establish interrupted brome, an extinct, naturalized grass, into a site in Oxfordshire. Since extinction is a natural process, should limited conservation resources be used to reintroduce species that have gone extinct? Should naturalized species that go extinct be reintroduced?

There are diverse approaches to species conservation. Some focus on conserving species in their natural habitats, such as nature reserves and national parks (*in situ* conservation), while others focus on conserving species away from their natural habitats, for example in gene banks and botanic gardens (*ex situ* conservation). Both *in situ* and *ex situ* approaches bring contrasting conservation philosophies into sharp focus. One emphasizes the maintenance of evolutionary processes, the other the conservation of evolutionary endpoints. Simplistically *ex situ* conservation is about endpoint preservation, while *in situ* conservation is about process preservation. The majority of current conservation programmes are concerned with conserving species, the evolutionary endpoints, rather than the processes that give rise to species. Of course, process conservation is a much more difficult, perhaps even an impossible, task.

The debate over how limited conservation resources should be spent is not over. It is indeed likely to become more significant in the future as environmental problems increase, more species need help and resources become more limited. There are no simple, objective answers to conservation priorities, but gene banks potentially provide a breathing space so informed decisions can be made; they are security blankets for *in situ* conservation. Gene banks are about spreading risk and increasing probabilities of species survival. They are global insurance policies against extinction and complement *in situ* approaches to conservation; they are not conservation panaceas. Gene banks are one element used to meet Target 8 of the Global Strategy for Plant Conservation that was adopted in 2002. Target 8 aimed to conserve 60 per cent of threatened plant species in accessible gene banks by the year 2010.

Gene banks may take many forms. The collection of living fruit trees at Brogdale in Kent is one type of gene bank. Another is the collection of potato tubers in the Commonwealth Potato Collection in Dundee. In other cases, collections of plants are maintained through tissue culture and grown inside bottles or held as bundles of cells at temperatures close to -196°C. However, seed banks have been used most widely for the long-term conservation of plant genetic material.

Humans have exploited the natural characteristics of seeds to ensure that plants are passed from generation to generation, migrate with human populations and, potentially, provide resilience to environmental change. However, for some plants the only means of conserving them is by growing them. There is no point in conserving apple varieties as seed since all seeds from a particular apple variety will be genetically different from the parental tree. Over two thousand apple cultivars in the UK are thus maintained and propagated as grafts; some of them have been propagated for hundreds of years. The extreme case of having to maintain plants as clone banks, with all the dangers of disease and 'acts of God' that this implies, is found in the cultivated banana, which is sterile. While plants disperse their genes in time and space, people who save plant genes for future generations have chosen to conserve seeds in a limited number of spaces.

The regions out of which wheat and other cereals emerged ten thousand years ago are today political powder kegs; few Western botanists have collected plants from them in the last two decades. Instead plant breeders interested in the plants from these regions have relied on accessions stored in gene banks, often decades ago. However, gene banks are threatened by neglect, underfunding and, in some parts of the world, war. The importance of conserving plants and their genetic resources for generations means that it has become a pressing global responsibility. In early 2007 the Norwegian Government announced that the Svalbard Global Seed Vault would be carved 120 m (130 yards) into the side of a mountain near the Norwegian city of Longyearbyen, some 130 m (140 yards) above current sea levels and about 1,000 km (620 miles) south of the North Pole.

The construction of a global seed depository had been discussed for nearly twenty-five years. However, it was not until the International Treaty on Plant Genetic Resources was signed that an acceptable international framework was in place to guide the activities of a global seed bank. Eventually the Norwegian government funded the construction of the Vault ($9 million) and will fund the annual maintenance costs. The Global Crop Diversity Trust funds the operation and management of the Vault. The location of the Vault was deliberate and based on realistic, worst-case climate scenarios. Close to the North Pole, natural permafrost ensures that the vault is kept cold, even if refrigeration fails. Placing the Vault deep underground means that temperature changes inside the Vault will be minimal, even if there are dramatic changes in outside air temperature. The height above sea level means that the Vault will be safe from sea level rises, even if both the Greenland and Antarctica ice sheets melt. The seeds are stored according to international seed bank standards, and – importantly – remain the property of the country or institution that deposited them and are freely available under the terms of the Treaty. Neither the Vault managers, the Norwegian government nor the Trust have any rights even to open the boxes in which the seeds are stored. The Vault opened in February 2008; after one year it contained some 400,000 duplicate seed accessions from twenty-five national and international institutions involved with the conservation of genetic resources.

The Svalbard Global Seed Vault is the latest addition to Nikolai Vavilov's legacy, a global network of gene banks established over the last five decades. Active decisions have been made and clear arguments put forward to create new gene banks. There are few good reasons for investing large amounts of resources in a gene bank just to make a political point. Once commitments have been made to build a gene bank, funders must be aware that they are playing the long game. To paraphrase an animal welfare slogan popular in the UK, 'A gene bank is forever, not just for a sound-bite.' Those responsible for maintaining Vavilov's seed bank during the Siege of Leningrad knew the significance of such collections (p. 185).

It is estimated that there are about six million seed accessions stored in about 1,300 seed banks worldwide. However, this number masks the fact that about 85 per cent of these are from a limited range of crop plants and their wild relatives. The vast majority of the world's flora remains unprotected, either in gene banks or in reserves. Many of the world's wild plants remain at severe risk of extinction. The emphasis of the Global Seed Vault on the conservation of food plants contrasts with the emphasis of the UK's Millennium Seed Bank on maximizing the conservation of diversity across all plants. Indeed, one of the targets of the Millennium Seed Bank was to collect 10 per cent of the world's flora by 2010. The Bank's success has meant that the target has been raised to 25 per cent of the world's flora by 2020. Of course, as the target is raised it becomes more and more difficult to reach. The first 25 per cent of the flora is much easier to bank than the last 25 per cent will be. Unfortunately we do not have the resources to conserve everything, so increasingly difficult conservation choices will have to be made.

Orphaned from their natural habitats, many plants are in danger of finding that their new habitats are hermetically sealed jars and packets inside gene banks. Species may inadvertently be selected for life in the gene bank rather than in the wild, just as early farmers selected plants that were better able to survive in a field under the hand of man rather than in the wild. The future of species in gene banks rests entirely with the managers and funders of such collections, the choices that they make and the information that they have about the samples in their custody. Gene bank managers must combine global vision with conservatism. In the words of the writer of the book of Proverbs, who may have been a gene bank manager himself, albeit of his family's crops: 'He that tilleth his land shall have plenty of bread: but he that followeth after vain persons shall have poverty enough.' Gene banks, like their financial cousins, should not be casinos, in which speculation and short-term gain take priority over responsible management and the minimization of risk. However, gene bank curators are pulled in three directions. Their natural desire to increase the size and quality of their collections must be balanced against the increased cost associated with a larger collection and the desire of funders to reduce or hold budgets static.

It is not enough for genes to be locked away in a hole in the side of a mountain; if a gene bank is to be useful, it must be actively managed. In addition, every sample in the gene bank must have a positive reason for being there, and must have the potential to earn its keep. Sound management decisions must therefore be based on sound scientific evidence.

Considerable research has been undertaken in the past two decades to determine how best to populate gene banks and how best to collect seeds from many types of plants. High quality additions to gene banks must have five characteristics. Obviously, the species from which the sample is taken must be correctly identified. However, correct identification can often be difficult to achieve in regions where the plants are poorly known. The plant part (usually seeds) put into a gene bank must be healthy and viable, the sample must be genetically representative and sufficiently large to be useful. There must also be adequate associated data. Unlike cultivated plants, the distributions of wild plants are often poorly known – even the rarities that attract the attention of the gene bank manager. Furthermore, basic biological knowledge such as when the species flowers, when it fruits and when its seeds are mature may be unknown. Some species, such as meranti trees in the Peninsular Malaysia, do not flower annually, while other species, such as European oaks, flower but do not produce large quantities of fruit annually. Wild species may also be difficult to access and the seeds physically difficult to collect, together with being short-lived and having diverse dormancy mechanisms.

Sampling for a gene bank should reproduce the genetic structure of the population or species in the field, which raises questions about the numbers of populations to be sampled per species, the number of individuals to be collected per population and the number of seeds to be harvested per individual. There is, of course, a paradox. The best sampling strategy for a seed collection would be based on knowledge of its genetic structure. However, to gain knowledge of genetic structure it is necessary to have sampled material effectively in the first place. The gene bank collector, especially of wild species, rarely has the luxury of sampling from a location twice. However, for some species these types of considerations are a luxury, and the conservation concern is such that saving some seed is better than saving none.

Once the material is in the seed bank, the question naturally arises whether seeds from the different individuals should be stored separately as half-sib families, or whether all the individuals should be mixed as a bulked seed collection. Further complications are raised when it is realized that seeds collected in different years represent different genetic samples of a population, since individuals vary from year to year in both their seed and pollen production. Given the potential significance of these issues for *in situ* conservation and effective seed banking, we know surprisingly little about the details of gene flow processes in natural populations of wild species. Since any sampling strategy will reduce the amount of genetic diversity of a sample, the trick is to try to minimize the reduction in genetic diversity of a sample.

Various guidelines have been published regarding the numbers of seeds that should be used. These guidelines suggest that a minimum of two thousand five hundred seeds should be collected per accession, since this allows for immediate use (propagation and germination trials), distribution for research and the anticipated loss in viability over time. The Millennium Seed Bank (48) aims to collect ten to twenty thousand seeds per species, so that seeds may be periodically tested over hundreds of years. The 'target minimum' may not be appropriate if the seeds are very large, they store poorly or removal of such a large amount of seed would affect the long-term viability of the species in the wild.

Ideally, every sample in a gene bank ought to have a 'passport' that states where, when and how it was collected, together with ecological and ethnobotanical information. The more passport information that exists, the more useful the sample is likely to be, since scientific value has as much to do with labels as with samples. In the same way that the scientific value of Kew's living plant collections was questioned in the 1960s, since many plants were from unknown, often cultivated, sources, poorly labelled samples in gene banks are likely to have little scientific value. Without detailed passport information, creating a gene bank is in danger of becoming little better than stamp collecting.

Piggy banks should be filled with strong currencies. However, we do not know what will be the strongest currencies in the future, therefore to minimize risks a mixture of currencies should be held. However, it is not enough just to keep putting material into the piggy bank; at some point, its contents must be liberated. Coins can be levered out using a knife or the bank could be smashed, but a wise investor will have ensured that there is a stopper in the bank's bottom for easy access. In the case of genetic 'piggy banks' it is important to know that the stored germplasm can be grown in the future. Can tubers in clone collections be propagated? Can plants be regenerated from tissue culture samples? Will seeds germinate after being stored?

Seeds that can be dried and stored in seed banks are described as orthodox, whereas desiccation-sensitive seeds are called recalcitrant. Up to 40 per cent of plants, for example avocado, litchi, mango and citrus, have been estimated to produce recalcitrant seed and cannot be stored for long periods. The ability to recognize plants with recalcitrant seeds quickly is obviously important, since making the wrong decisions about gene bank technologies could be disastrous for the conservation of a species. Seed storage lives also vary considerably across species, even when the seeds are stored under ideal conditions. Awareness of seed shelf lives will affect how collections are managed. Even if seeds can survive in the seed bank, at some point it will be necessary to grow the whole plant and therefore the seed must germinate. Seed banks have been particularly effective at investigating effective germination and propagation protocols, especially in the last two decades. Many seed banks have research programmes in place with the objective of understanding how the processes of seed development work, and how factors such as ecology, climate and seed structure affect seed development.

Comparisons among gene banks merely based on numbers of accessions assume that all accessions have equal value. In practice this cannot be the case, since there is redundancy in gene banks. Some accessions will be very poorly documented, others will be duplicates of accessions in other collections, others will be of species extinct or endangered in the wild and yet others will contain agriculturally important genes. Given the costs of maintaining gene banks, this has led some managers of gene banks to ask how redundancy can best be detected and eliminated in the collections for which they are responsible. Such matters raise profound questions about how a gene bank manager makes the decision to discard a sample – and who takes responsibility for that decision, as there are no absolutes, only probabilities that a sample will be of no future use. Genetic resources are too important for these decisions to be made on a whim.

One of the themes of this book has been the decline of biodiversity in agricultural landscapes over the past century. Such losses include the disappearance of traditional crop varieties and the loss of weed species. As we have seen, seed genetic resources of the main agricultural plants have been the focus of intense interest by an international network of seed banks. However, this is only the start. We appear to have been lucky with the conservation of the main agricultural plants and their wild relatives, though what was destroyed through ignorance and dogma is impossible to determine. We are unlikely to be so lucky with the altogether more daunting task of conserving wild plant species with little or no current economic value.

The history of our involvement with seeds will continue. We have entered a new phase in which we can no longer claim ignorance of the problems we, and the plants upon which we depend, face. The decisions we make now will have long-lasting consequences for future generations. Only those generations will be able to judge whether our enormous investment in seed banks has been worthwhile. Just as the unconscious decisions made by peasant farmers four hundred generations ago had consequences for today's generation, so our conscious decisions may have important, and unforeseen, consequences for the genetic resources available to future generations. It is to be hoped that we will make our choices rationally and equitably for the benefit of the whole of humankind, rather than for the political, social or economic benefit of the few.

Further Reading

Baskin, C. C., and J. M. Baskin, *Seeds: Ecology, Biogeography, and Evolution of Dormancy and Germination*, San Diego: Academic Press, 2001

Bioversity International, http://www.bioversityinternational.org/bioversity_international_homepage.html

Briggs, D., and S. M. Walters, *Plant Variation and Evolution*, Cambridge: Cambridge University Press, 1997

Darwin, F. (ed.), *The Life and Letters of Charles Darwin, including an autobiographical chapter*, 3 vols, London: John Murray, 1887

Frankel, O. H., A. H. Brown and J. J. Burdon, *The Conservation of Plant Biodiversity*, Cambridge: Cambridge University Press, 1995

Global Strategy for Plant Conservation, http://www.plants2010.org/

Harlan, J. R., *Crops and Man*, Madison, WI: American Society of Agronomy, 1992

Hawkes, J. G., N. Maxted and B. V. Ford-Lloyd, *The ex situ Conservation of Plant Genetic Resources*, Dordrecht: Kluwer Academic Press, 2000

Heiser, C. B., *Seed to Civilization: The Story of Food*, Cambridge, Mass and London: Harvard University Press, 1990

Hobhouse, H., *Seeds of Wealth: Four Plants that Made Men Rich*, London: Pan, 2004

Hobhouse, H., *Seeds of Change: Six Plants that Transformed Mankind*, London: Shoemaker & Hoard, 2006

Howes, C., *The Spice of Life: Biodiversity and the Extinction Crisis*, London: Blandford, 1997

Intergovernmental Panel on Climate Change, http://www.ipcc.ch/

Juma, C., *The Gene Hunters. Biotechnology and the Scramble for Seeds*, Princeton, New Jersey: Princeton University Press, 1989

Kesseler, R., and W. Stuppy, *Seeds: Time Capsules of Life*, London: Papadakis, 2009

Kingsbury, N., *Hybrid: The History and Science of Plant Breeding*, Chicago and London: University of Chicago Press, 2009

Maxted, N., B. V. Ford-Lloyd and J. G. Hawkes, *Plant Genetic Conservation: The in situ Approach*, London: Chapman & Hall, 1997

Mazoyer, M., and L. Roudart, *A History of World Agriculture from the Neolithic Age to the Current Crisis*, London: Earthscan, 2006

Murphy, D. J., *People, Plants and Genes: The Story of Crops and Humanity*, Oxford: Oxford University Press, 2007

Nabhan, G. P., *Where our Food Comes From: Retracing Nikolay Vavilov's Quest to End Famine*, Washington, DC: Island Press, 2008

Niklas, K. J., *The Evolutionary Biology of Plants*, Chicago and London: University of Chicago Press, 1997

Proctor, M., P. Yeo and A. Lack, *The Natural History of Pollination*, London: Harper Collins, 1996

Raven, P. H., R. F. Evert and S. E. Eichorn, *The Biology of Plants*, New York: Worth, 1998

Silvertown, J., *An Orchard Invisible: A Natural History of Seeds*, Chicago and London: University of Chicago Press, 2009

Smith, B. D., *The Emergence of Agriculture*, New York: Scientific American Library, 1998

Smith, R. D., J. B. Dickie, S. H. Linington, H. W. Pritchard and R. J. Probert, *Seed Conservation: Turning Science into Practice*, Kew: Royal Botanic Gardens, 2003

Stocks, C., *Forgotten Fruits: A Guide to Britain's Traditional Fruit and Vegetables*, London: Random House, 2008

ten Kate, K., and S. A. Laird, *The Commercial Use of Biodiversity: Access to Genetic Resources and Benefit-Sharing*, London: Earthscan, 1999

Tudge, C., *The Engineer in the Garden. Genes and Genetics: From the Idea of Heredity to the Creation of Life*, London: Jonathan Cape, 1993

Vavilov, N. I., *Origin and Geography of Cultivated Plants*, Cambridge: Cambridge University Press, 2009

Wilson, E. O., *The Future of Life*, London: Little, Brown, 2002

Zohary, D., and M. Hopf, *Domestication of Plants in the Old World*, Oxford: Oxford University Press, 2000

Picture Credits

Acknowledgments

This book grew out of my work at the Royal
Botanic Gardens, Kew and Wakehurst Place.
I would like to thank the friends and colleagues
there whose assistance and support were
crucial to the conduct of experiments on seed
germination and the early development of the
Seed Bank – particularly Anne Alexander,
Stephanie Cox (*née* Ebbels), David Fox, Roger
Smith, Robert Sanderson, Pat Newman, Doreen
McNamara, Jane Pescheira and Ted Ormerod.
I am also grateful to the gardeners in the Living
Collections Department at Kew, without whose
efforts there would never have been a seed bank
of any kind – notably Dick Shaw, George Brown,
Jim Mateer, Malcolm Latto, Tony Lyall, Pauline
Anthony, Ann Reekie, Margaret Freeman and
Doreen Lower.

Finally, I would like to express my gratitude to
Linda Hurcombe for her encouragement and
support, and above all for persuading me that
someone, somewhere, might enjoy reading
this book.

Index